AF255705

NEW

PLANE GEOMETRY

BY

WOOSTER WOODRUFF BEMAN

PROFESSOR OF MATHEMATICS IN THE UNIVERSITY OF MICHIGAN

AND

DAVID EUGENE SMITH

PRINCIPAL OF THE STATE NORMAL SCHOOL AT BROCKPORT
NEW YORK

BOSTON, U.S.A.

GINN & COMPANY, PUBLISHERS

The Athenæum Press

1899

PREFACE.

THE demand of a large number of schools for a book confined to plane geometry leads the authors to issue this edition of the first part of their "New Plane and Solid Geometry." In offering it to the profession an explanation of its distinctive features may be of service.

It is sometimes asserted that we should break away from the formal proofs of Euclid and Legendre and lead the student to independent discovery, and so we find text-books that give no proofs, others that give hints of the demonstrations, and still others that draw out the demonstration by a series of questions which, being capable of answer in only one way, merely conceal the Euclidean proof. But, after all, the experience of the world has been that the best results are secured by setting forth a minimum of formal proofs as models, and a maximum of unsolved or unproved propositions as exercises. This plan has been followed by the authors, and the success of the first edition has abundantly justified their action.

There is a growing belief among teachers that such of the notions of modern geometry as materially simplify the ancient should find place in our elementary text-books. With this belief the authors are entirely in sympathy. Accordingly they have not hesitated to introduce the ideas of one-to-one correspondence, of anti-parallels, of negative magnitudes, of general figures, of similarity of point.systems, and such other concepts as are of real value in the early study of the science. All this has been done in a conservative way, and such material as the first edition (1895) showed to be at all questionable has been omitted from the present revision.

Within comparatively recent years the question of methods of attack has interested several leading writers. Whatever has been found to be usable in elementary work the authors have inserted where it will prove of most value. To allow the student to grope in the dark in his efforts to discover a proof, is such a pedagogical mistake that this innovation in American text-books has been generally welcomed. Upon this point the authors have freely drawn from the works of Petersen of Denmark, and of Rouché and de Comberousse of France, and from the excellent treatise recently published by Hadamard (Paris, 1898).

With this introduction of modern concepts has necessarily come the use of certain terms and symbols which may not generally be recognized by teachers. These have, however, been chosen only after most conservative thought. None is new in the mathematical world, and all are recognized by the leading writers of the present time. They certainly deserve place in our elementary treatises on the ground of exactness, of simplicity, and of their general usage in mathematical literature.

The historical notes of the first edition have been retained, it being the general consensus of opinion that they add materially to the interest in the work. For teachers who desire a brief but scholarly treatment of the subject the authors refer to their translation of Fink's "History of Elementary Mathematics" (Chicago, The Open Court Publishing Co., 1899). For the limitations of elementary geometry, the impossibility of trisecting an angle, squaring a circle, etc., teachers should read the authors' translation of Klein's valuable work, "Famous Problems of Elementary Geometry" (Boston, Ginn & Company).

It is impossible to make complete acknowledgment of the helps that have been used. The leading European text-books have been constantly at hand. Special reference, however, is due to such standard works as those of Henrici and Treutlein, "Lehrbuch der Elementar-Geometrie," the French writers already mentioned, and the noteworthy contributions of the recent Italian school represented by Faifofer, by Socci and Tolomei, and by Lazzeri and Bassani.

Teachers are urged to consider the following suggestions in using the book:

1. Make haste slowly at the beginning of each book.

2. Never attempt to give all of the exercises to any class. From a third to a half, selected by the teacher, should suffice.

3. Require frequent written work, thus training the eye, the hand, and the logical faculty together. The authors' Geometry Tablet (Ginn & Company) is recommended for this work.

W. W. BEMAN, Ann Arbor, Mich.
D. E. SMITH, Brockport, N. Y.

August 15, 1899.

CONTENTS.

INTRODUCTION.

Book I. — Rectilinear Figures.

Book II. — Equality of Polygons.

Book III. — Circles.

APPENDIX TO BOOK III.

BOOK IV. — RATIO AND PROPORTION.

BOOK V. — MENSURATION OF PLANE FIGURES.
REGULAR POLYGONS AND THE CIRCLE.

APPENDIX TO PLANE GEOMETRY.

TABLES.

PLANE AND SOLID GEOMETRY.

PLANE GEOMETRY.

INTRODUCTION.

1. ELEMENTARY DEFINITIONS.

1. In **Arithmetic** the student has considered the science of *numbers*, and has found, for example, that a number which ends in 5 or 0 is divisible by 5.

In **Algebra** he has studied, among other things, the *equation*, and has found that if $\frac{1}{2}x - 1 = 5$, x must equal 12.

In **Geometry** he is to study *form*, and he will find, for example, that two triangles must necessarily be equal if the three sides of the one are respectively equal to the three sides of the other.

Before beginning the subject, however, there are certain terms which, although familiar, are used with such exactness as to require careful explanation. These terms are *solid, surface, line, angle* (with various kinds of each), and *point*. As with most elementary mathematical terms, such as *number, space*, etc., it is difficult to give them simple and satisfactory definition. Explanations can, however, be given which will lead the student to a reasonable understanding of them.

2. The space with which we are familiar and in which we live is evidently divisible. Any limited portion of space is called a **solid.**

In geometry no attention is given to the substance of which the solid is composed. It may be water, or iron, or air, or wood, or it may be a vacuum. Indeed, geometry considers only the *space occupied* by the substance. This space is called a *geometric solid*, or simply a *solid*, while the substance is called a *physical solid*. Thus, a ball is a physical solid; the space which the ball occupies is a geometric solid.

3. That which separates one part of space from an adjoining part is called a **surface**. So we speak of the surface of a ball, the surface of the earth, etc.

4. Every surface is divisible. That which separates one part of a surface from an adjoining part is called a **line**.

5. Every line is divisible. That which separates one part of a line from an adjoining part is called a **point.**

A point is not divisible.

Thus, in the figure the surface of the block separates the space occupied by the block from all the rest of space. This surface is divisible in 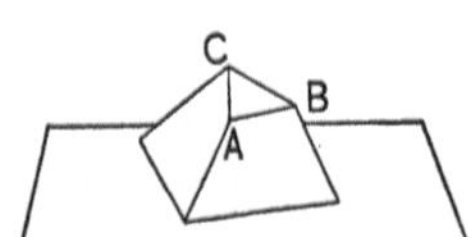 many ways; for example, it is divided into two parts by the line passing from *A* through *B* and *C* and back to *A*. This line is divisible in many ways; for example, it is separated into three parts by the points *A*, *B*, *C*. In the case of a line that returns into itself, — *i.e.* a *closed* line, like the one just mentioned, — *two* points are necessary completely to separate one part from the other.

It is impossible to draw mechanically a geometric line. A chalk mark, a thread, a fine wire, an ink mark, are all very thin physical solids used to *represent* lines; for this purpose they are very helpful. So, too, a dot may be used to *represent* a point, and a sheet of paper may be used to *represent* a surface, although each is really a physical solid.

6. The preceding definitions start from the *solid* and take the *surface, line,* and *point* in order. It is also possible to start with the *point* and proceed in reverse order.

The **point** is the simplest geometric concept; it has position, but not magnitude.

A moving point describes a **line**.

This may be *represented* by a pencil point moving on a piece of paper.

A moving line describes, in general, a **surface**.

This may be *represented* by a crayon lying flat against the blackboard, and moving sidewise. How may a line move so as not to describe a surface ?

A moving surface describes, in general, a **solid**.

Thus, the surface of a glass of water, as it moves upward, may be said to describe a solid. How may a surface move so as not to describe a solid ?

7. Through two points any number of lines may be imagined to pass.

For example, through the points P_1, P_2 (read " P-one, P-two ") the lines q, r, s may be imagined to pass.

A straight line is a line which is determined by any two of its points.

In the figure, s represents a straight line, for, given the points P_1, P_2 on the line, its position is fixed ; it is *determined*.

But q and r do not represent straight lines, because P_1 and P_2 do not determine them.

The word *line*, used alone, is to be understood to refer to a straight line.

The expression *straight line* is used to mean both an unlimited straight line and a portion of such a line. In case of doubt, *line-segment*, or merely *segment*, is used to mean a limited straight line.

As has been seen, a point is usually named by some capital letter. A segment is usually named by naming its end points, or by a single small letter.

In the annexed figure, AB, AC, BC, and o are marked off.

Two segments are said to be *equal* when they can be made to coincide.

8. If three points, *A, B, C*, are taken in order on a line, as in the preceding figure, then the line-segment *AC* is called the **sum** of the line-segments *AB* and *BC*, and *AB* is called the **difference** between *AC* and *BC*.

9. If a point divides a line-segment into two equal segments, it is said to **bisect** the line-segment and is called its **mid-point.**

A line is easily bisected by the use of a straight-edge and compasses, thus :

With centers *A* and *B*, and equal radii, describe arcs intersecting at *P* and *P'*.

Draw *PP'.'* This bisects *AB*.

The proof of this fact is given later.

10. If a segment is drawn out to greater length, it is said to be **produced.**

To produce *AB* means to extend it through *B*, toward *C*, in the second figure in § 7. To produce *BA* means to extend it through *A*, away from *B*.

11. A line not straight, but made up of straight lines, is called a **broken line.**

12. Through three points, not in a straight line, any number of surfaces may be imagined to pass.

For example, through the points *A, B, C* the surfaces *P* and *S* may be imagined to pass.

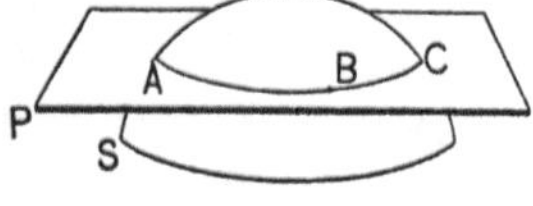

A **plane surface** (also called a *plane*) is a surface which is determined by any three of its points not in a straight line.

In the figure, *P* represents a plane, for it is *determined* by the points *A, B, C.* But *S* does not represent such a surface.

A plane is indefinite in extent unless the contrary is stated. To *produce* it means to extend it in length or breadth.

13. If two lines proceed from a point, they are said to form an **angle,** the lines being called the **arms,** and the point the **vertex,** of that angle.

The *size of the angle* is independent of the length of the arms; the size depends merely upon the amount of turning necessary to pass from one arm to the other.

The methods of naming an angle will be seen from the annexed figures. It is convenient to letter an angle around the vertex, as indicated by the arrows, that is, opposite to the course of clock-hands, or *counter-clockwise.*

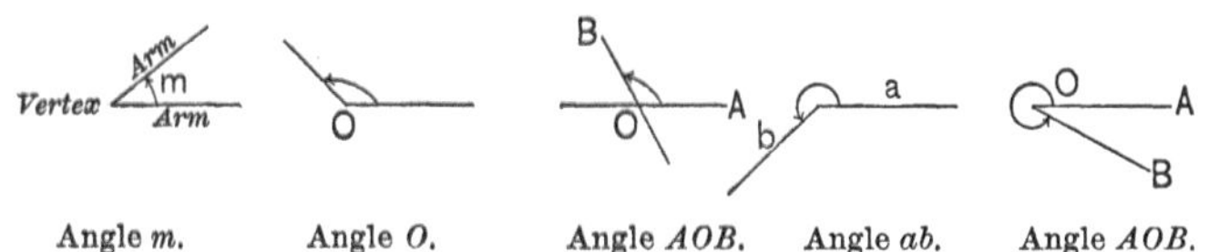

Angle *m*.   Angle *O*.   Angle *AOB*.   Angle *ab*.   Angle *AOB*.

A line proceeding from the vertex, turning about it counter-clockwise from the first arm to the second, is said to *turn through the angle, the angle being greater as the amount of turning is greater.*

14. If the two arms of an angle lie in the same straight line on opposite sides of the vertex, a **straight angle** is said to be formed. If the angle still further increases, until the moving arm has performed a complete revolution, thus passing through two straight angles, a **perigon** is said to be formed.

For practical purposes angles are measured in degrees, minutes, and seconds. A perigon is said to contain 360°.

AOB, a straight angle.
BOA, a straight angle.

A perigon, or angle of 360°.

In general, if two lines are drawn from *O*, two angles, each less than a perigon, are formed. Of these the smaller is always to be understood if "the angle at *O*" is mentioned, unless the contrary is stated.

15. If a line turns through an angle, all points or line-segments through which it passes in its turning, except the vertex, are said to be *within the angle*. Other points or lines are either on the arms or *without the angle*.

16. Two angles, *ab*, *a'b'*, are said to be *equal* when, without changing the relative position of *a* and *b*, angle *ab* may be placed so that *a* lies along *a'*, and *b* along *b'*.

This equality is tested by placing one angle on the other, the vertices coinciding. Then if the arms can be made to coincide, the angles are equal, otherwise not.

17. If three lines, *OA*, *OB*, *OC*, proceed from a common point *O*, *OB* lying within the angle *AOC*, then angles *AOB* and *BOC* are called **adjacent angles.** Angle *AOC* is called the *sum of the angles AOB, BOC*. Either of the adjacent angles is called the *difference* between angle *AOC* and the other of the adjacent angles.

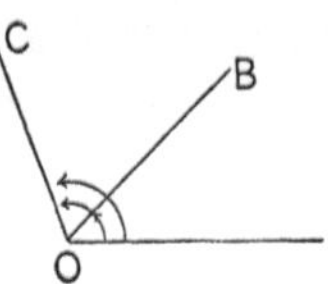

As two angles may be added, so several may be added.

18. If a line divides an angle into two equal angles, it is said to *bisect* the angle and is called its *bisector*.

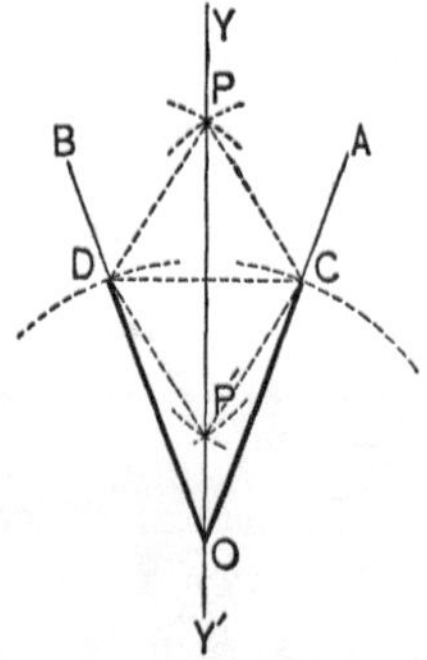

In the annexed figure, if angle *AOY* equals angle *YOB*, then *OY* is the bisector of angle *AOB*.

And, in general, to *bisect* any magnitude means to divide it into two equal parts.

An angle is easily bisected by the use of a straight-edge and compasses, thus:

If *AOB* is the given angle, mark off with the compasses *OC* equal to *OD*.

Then with *C* and *D* as centers and *CD* as a radius draw two arcs intersecting at *P* and *P'*.

The line joining *P* or *P'* with *O* is the required bisector. The **proof** of this fact is given later.

19. A **right angle** is half of a straight angle.

It follows from this definition that *the sum of two right angles is a straight angle;* and from the definitions of a straight angle and of a perigon, *that the sum of two straight angles, or of four right angles, is a perigon.*

It also follows that a straight angle contains 180° and a right angle contains 90°.

20. If two lines meet and form a right angle, each line is said to be **perpendicular** to the other.

Each is also spoken of as *a* perpendicular to the other. Thus, in the preceding figure, BO is perpendicular to CA, or is *a* perpendicular to CA. The segment PO is called *the* perpendicular from P to CA, since it will presently be proved that it is unique; that is, that there is one and only one perpendicular. O is called the *foot* of that perpendicular.

The word *unique,* meaning *one and only one,* is frequently used in mathematics.

A line is easily drawn perpendicular to another line by the use of a straight-edge and compasses. This is seen in the figure in § 9, where PP' is perpendicular to AB.

21. An angle less than a right angle is said to be **acute;** one greater than a right angle but less than a straight angle is said to be **obtuse;** one greater than a straight angle but less than a perigon is said to be **reflex** or **convex.**

22. Two lines which form an acute, obtuse, or reflex angle are said to be **oblique** to each other.

Acute, obtuse, and reflex angles are classed under the general term *oblique angles.*

The meaning of the expressions *oblique lines, an* oblique, *foot* of an oblique, will be understood from § 20.

Draw a figure representing acute, obtuse, and reflex angles, oblique lines, an oblique from P to CA, the foot of an oblique.

23. Two angles are said to be **complements** of each other if their sum is a right angle. Two angles are said to be **supplements** of each other if their sum is à straight angle. Two angles are said to be **conjugates** of each other if their sum is a perigon.

If one angle is the complement of another, the two angles are said to be *complemental* or *complementary*. Similarly, if one angle is the supplement of another, the two angles are said to be *supplemental* or *supplementary*.

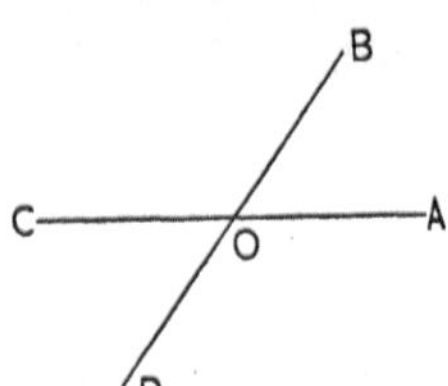

In the annexed figure, angles *A OB* and *BOC* are supplemental, also angles *BOC* and *COD*, etc.

24. If two lines, *CA*, *DB*, intersect at *O*, as in the above figure, the angles *AOB* and *COD* are called **vertical** or **opposite** angles; also the angles *BOC* and *DOA*.

Exercises. 1. How many degrees in a right angle? How many minutes? How many seconds?

2. What is the complement of one-half of a right angle? of one-fourth?

3. How many degrees in the supplement of an angle of (a) 75°? (b) 90°? (c) 150°? (d) 179°?

4. Also in the complement of an angle of (a) 75°? (b) 1°? (c) 89°? (d) 45°? (e) 90°? (f) 0°?

5. Also in the conjugate of an angle of (a) 270°? (b) 180°? (c) 359°? (d) 90°? (e) 1°? (f) 360°?

6. Draw a figure showing that two straight lines determine one point; also one showing that three straight lines determine, in general, three points.

7. How many degrees in each of the two conjugate angles which the hour and minute hands of a clock form at 4 o'clock?

8. If six lines, proceeding from a point, divide a perigon into six equal angles, express one of those angles (a) in degrees, (b) as a fraction of a right angle, (c) as a fraction of a straight angle.

2. THE DEMONSTRATIONS OF GEOMETRY.

25. The **object of geometry** is the investigation of *truths* concerning combinations of lines and points, and of the methods of making certain *constructions* from lines and points.

26. A **proposition** is a statement of either a truth to be demonstrated or a construction to be made.

. For example, geometry investigates this proposition: If two lines intersect, the vertical angles are equal. It also investigates the methods of drawing a line perpendicular to another line, and various other propositions requiring some construction.

Propositions are divided into two classes — *theorems* and *problems.*

A **theorem** is a statement of a geometric *truth* to be demonstrated.

A **problem** is a statement of a geometric *construction* to be made.

For example: Theorem, *If two lines intersect, the vertical angles are equal.* — Problem, *Required through a point in a line to draw a perpendicular to that line.*

27. There are a few *geometric* statements so obvious that the truth of them may be taken for granted, and a few geometric operations so simple that it may be assumed that they can be performed. Such a statement, or the claim to perform such an operation, is called a **postulate.**

The geometric operations thus assumed require the use of the straight-edge and compasses. *The straight-edge and the compasses are the only instruments recognized in elementary geometry.*

The postulates used in this work are set forth from time to time as required. At present three general classes suffice, as follows:

28. **Postulates of the Straight Line.**

1. *Two points determine a straight line.*
This follows from the definition.

2. *Two straight lines in a plane determine a point.*

3. *A straight line may be drawn and revolved about one of its points as a center so as to include any assigned point in space.*

4. *A straight line-segment may be produced.*

5. *A straight line is divided into two parts by any one of its points.*

29. **Postulates of the Plane.**

1. *Three points not in a straight line determine a plane.*
This follows from the definition.

2. *A straight line through two points in a plane lies wholly in the plane.*

Thus, if part of a straight line lies in an unlimited plane blackboard, the whole line lies in the blackboard.

3. *A plane may be passed through a straight line and revolved about it so as to include any assigned point in space.*

4. *A portion of a plane may be produced.*

5. *A plane is divided into two parts by any one of its straight lines, and space is divided into two parts by any plane.*

30. **Postulate of Angles.**

All straight angles are equal.

31. There are also a number of simple statements, *of a general nature*, so obvious that the truth of them may be taken for granted. These are called **axioms**.

The following are the axioms most frequently used in geometry, and they are so important that they should be learned by number.

32. Axioms.

1. *Things which are equal to the same thing, or to equal things, are equal to each other.*

That is, (1) if $A = B$, and $C = B$, then $A = C$.   Or, (2) if $A = B$, and $B = C$, and $C = D$, then $A = D$.

2. *If equals are added to equals, the sums are equal.*

That is, if $A = B$, and if $C = D$, then $A + C = B + D$.

3. *If equals are subtracted from equals, the remainders are equal.*

That is, if $A = B$, and if $C = D$, then $A - C = B - D$.

4. *If equals are added to unequals, the sums are unequal in the same sense.*

That is, if $A = B$, and if C is greater than D, then $A + C$ is greater than $B + D$.

5. *If equals are subtracted from unequals, the remainders are unequal in the same sense.*

That is, if $A = B$, and if C is greater than D, then $C - A$ is greater than $D - B$.

6. *If equals are multiplied by equals, the products are equal.*

That is, if $A = B$, and m is any number, then $mA = mB$.

7. *If equals are divided by equals, the quotients are equal.*

That is, as in axiom 6, $\dfrac{A}{m} = \dfrac{B}{m}$.   It will be seen that axiom 6 covers axiom 7, for m may be a fraction.

8. *The whole is greater than any of its parts, and equals the sum of all its parts.*

The latter part of this axiom is merely the definition of *whole*.

9. *If three magnitudes are so related that the first is greater than the second, while the second is greater than, or equal to, the third, then the first is greater than the third.*

E.g. if A is greater than B, and if B is greater than, or equal to, C, then A is greater than C.

33.   SYMBOLS AND ABBREVIATIONS.

The following are used in this work, and are inserted here merely for reference, and not for memorizing:

e.g.	Latin, *exempli gratia*, for example.
i.e.	Latin, *id est*, that is.
∵	since.
∴	therefore.
pt., pts.	point, points.
rt.	right.
st.	straight.
ax.	axiom.
post.	postulate.
def.	definition.
prop.	proposition.
th.	theorem.
pr.	problem.
cor.	corollary.
subst.	substitution.
prel.	preliminary.
const.	construction.
ppd.	parallelepiped.
⌒	arc.
⊙, Ⓢ	circle, circles.
△, △s	triangle, triangles.
□, ☐s	square, squares.
▭, ▭s	rectangle, rectangles.
▱, ▱s	parallelogram, parallelograms.
∠, ∠s	angle, angles.
+	plus, increased by.
−	minus, diminished by.
×, · ,	and absence of sign, denote multiplication.

$\div$, /, :, and fractional form, denote division.

= is equal, or equivalent, to.

≡ is identical with, as $AB \equiv AB$, or coincides with.

≅ is congruent to.

∽ is similar to.

≐ approaches as a limit.

> is greater than.

< is less than.

≠ is not equal to, *i.e.* > or <.

≯ is not greater than, *i.e.* = or <.

≮ is not less than, *i.e.* = or >.

⊥ is perpendicular to, or a perpendicular.

‖ is parallel to, or a parallel.

..... and so on.

The above take the plural also; thus, = means *are* equal, as well as *is* equal.

The manner of reading some of the familiar symbols is suggested, as follows:

P', P-prime; P'', P-second; P''', P-third, etc.

P_1, P-one; P_2, P-two, etc.

$A'B'$, A-prime B-prime, etc.

A_1', A-one-prime, etc.

References to preceding propositions are made by book and proposition thus, I, prop. IV; if the first Roman numeral is omitted, the proposition is in the current book. Section references are also used.

Other simple abbreviations are occasionally used, but they will be easily understood.

3. PRELIMINARY PROPOSITIONS.

34. The following theorems are designed to show to the beginner the nature of a geometric proof, and to lead him by easy steps to appreciate the logic of geometry. Some of them might properly have been incorporated in Book I, and others might have been omitted altogether; but they form a group of simple propositions which lead the student up to the more difficult work of geometry, and for that reason they are inserted here. The student and the teacher are advised to proceed slowly until the logic of the subject is understood, and under no circumstances to allow mere memorizing of the proofs.

Proposition I.

35. Theorem. *All right angles are equal.*

Suggestion. The only angles of whose equality we are thus far assured are straight angles. Hence in some way we must base our proof of this theorem on the postulate of angles, which asserts this fact. We then consider how a right angle is related to a straight angle, and the proof is at once suggested.

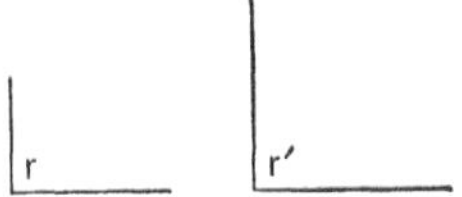

Given any two right angles, r, r'.

To prove that $r = r'$.

Proof. 1. r and r' are halves of straight angles.   Def. rt. $\angle$
      (§ 19. A right angle is half of a straight angle.)

  2. All straight angles are equal.   § 30

  3. ∴ all right angles, and hence r and r', are equal.

Ax. 7

(If equals are divided by equals, the quotients are equal.)

PROPOSITION II.

36. Theorem. *At a given point in a given line not more than one perpendicular can be drawn to that line in the same plane.*

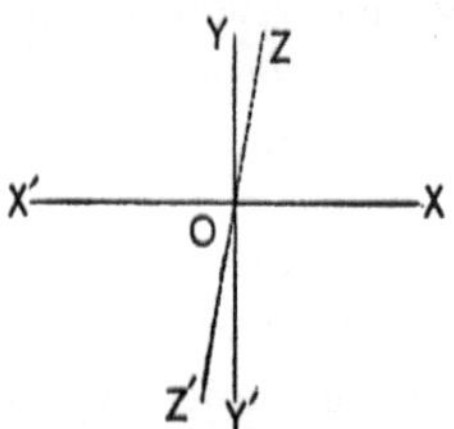

Given $YY' \perp XX'$ at O.

To prove that no other perpendicular can be drawn to XX', at O, in the same plane.

Proof. 1. Suppose that another $\perp$, ZZ', could be drawn.

2. Then $\angle XOZ$ would be a rt. $\angle$.   Def. $\perp$

(If two lines meet and form a rt. $\angle$, each is said to be $\perp$ to the other.)

3. But $\angle XOY$ is a rt. $\angle$.   Given; def. $\perp$, § 20

(For it is given that $YY' \perp XX'$, and the def. of a $\perp$ is given in step 2.)

4. $\therefore \angle XOY$ would equal $\angle XOZ$.   Prop. I

(All right angles are equal.)

5. But this is impossible.   Ax. 8

(The whole is greater than any of its parts, etc.)

6. $\therefore$ the supposition of step 1 is absurd, and a second perpendicular is impossible.

NOTE. In prop. I we proved directly from the definition of straight angle that all right angles are equal. In prop. II a different method of proof is followed. We have here supposed that the theorem is false and have shown that this supposition is absurd. Such proofs have long been known by the name "*reductio ad absurdum*," a reduction to an absurdity. They are also called *indirect* proofs.

Proposition III.

37. Theorem.   *The complements of equal angles are equal.*

Suggestion.   Three lines of proof may present themselves.   We may base our proof on the equality of straight angles, as we did in prop. I, or we may take an indirect proof as in prop. II, beginning by supposing the theorem false and showing the absurdity of this supposition, or we may base the proof on prop. I.   Since the complements suggest right angles, which of the three methods would it probably be best to follow ?

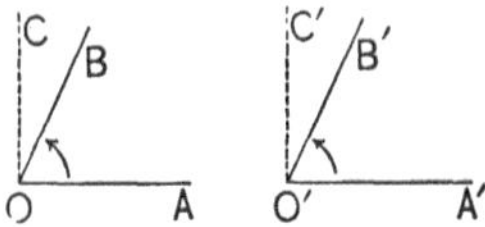

Given   two equal $\angle$s, AOB, $A'O'B'$, and their complements, BOC, $B'O'C'$, respectively.

To prove   that $\angle BOC = \angle B'O'C'$.

Proof. 1. $\angle$s AOC and $A'O'C'$ are rt. $\angle$s.   Def. compl.
       (§ 23.   Two $\angle$s are said to be complements if their sum is a rt. $\angle$.)

   2. $\therefore \angle AOC = \angle A'O'C'$.   Prop. I
              (All right angles are equal.)

   3. But $\angle AOB = \angle A'O'B'$.   Given

   4. $\therefore \angle BOC = \angle B'O'C'$.   Ax. 3

(If equals are subtracted from equals, the remainders are equal.)

Proposition IV.

38. Theorem.   *The supplements of. equal angles are equal.*

Let the student draw the figure and give the proof after the manner of prop. III.   Use only four steps in the proof.

Given

To prove

Proof.

PROPOSITION V.

39. Theorem. *The conjugates of equal angles are equal.*

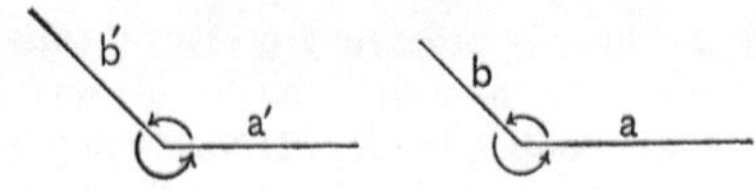

Given   two equal angles, ab, $a'b'$.

To prove that $\angle ba = \angle b'a'$.

Proof. 1. The given $\angle\!\!\!\angle$ may be so placed that a lies along a',
and b along b'.   Def. equal $\angle\!\!\!\angle$

(§ 16. Two $\angle\!\!\!\angle$, ab, $a'b'$, are said to be equal when $\angle ab$ can be placed
so that a lies along a', and b along b'.)

2. But then $\angle ba$ must equal $\angle b'a'$.   Def. equal $\angle\!\!\!\angle$

PROPOSITION VI.

40. Theorem. *If two lines cut each other, the vertical
angles are equal.*

SUGGESTION. After examining the figure the student might say that
because $\angle a + \angle b =$ st. $\angle$, and $\angle b + \angle a' =$ st. $\angle$, $\therefore \angle a + \angle b = \angle b + \angle a'$, and then subtract $\angle b$ from these equals; or he might say that
$\angle a = \angle a'$ because each is the supplement of $\angle b$. He should always feel
encouraged to try various proofs, selecting the shortest and the clearest.
Does the following proof meet these requirements ?

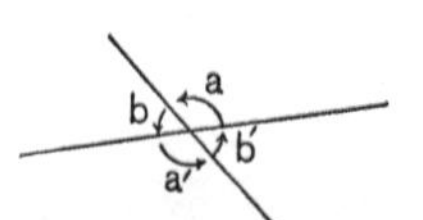

Given   two lines cutting each other,
forming two pairs of opposite
angles, a, a', and b, b'.

To prove that $\angle a = \angle a'$.

Proof. 1. $\angle a$ and $\angle a'$ are supplements of $\angle b$.   Def. suppl.

(§ 23. Two $\angle\!\!\!\angle$ are said to be supplements if their sum is a st. $\angle$.)

2. $\therefore \angle a = \angle a'$.   Prop. IV

(The supplements of equal angles are equal.)

Proposition VII.

41. Theorem. *A line-segment can be bisected in only one point.*

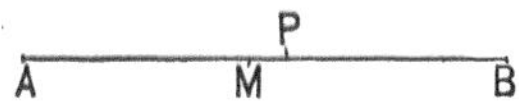

Given   a line-segment AB, bisected at M.

To prove   that there is no other point of bisection.

Proof.   1. Suppose another point of bisection exists, as P, between M and B.

2. Then since AM and AP are both halves of AB, they are equal.   Ax. 7
   (State ax. 7.)

3. But this is impossible, for AM is part of AP.   Ax. 8
   (State ax. 8.)

4. ∴ the supposition that there is a second point of bisection is absurd.

   (Another *reductio ad absurdum*, as in prop. II.)

Proposition VIII.

42. Theorem. *An angle can be bisected by only one line.*
(The student may prove this after the manner of prop. VII.)

Exercises. 9. Of two supplemental angles, a and b, (a) suppose $a = 2\,b$, how many degrees in each? (b) suppose $a = 3\,b$, how many?

10. How many straight lines are, in general, determined by three points? by four? (The points in the same plane.)

11. If of five angles, a, b, c, d, e, whose sum is a perigon, $a = 20°$, $b = 30°$, $c = 40°$, $d = 50°$, how many degrees in e?

12. Of three angles whose sum is a perigon, the first is twice the second, and the second three times the third; how many degrees in each?

PROPOSITION IX.

43. Theorem.	*The bisectors of two adjacent angles formed by one line cutting another are perpendicular to each other.*

SUGGESTION. Considering the figure, we see that to prove $OA \perp OB$ we must show that $\angle AOB$ is a rt. $\angle$. Now the only way that we have as yet of showing an angle to be a right angle is to show that it is half of a straight angle. But evidently $\angle AOY$ is half of $\angle XOY$, because $\angle XOY$ is bisected; similarly, $\angle YOB$ is half of $\angle YOX'$, and this suggests the following proof.

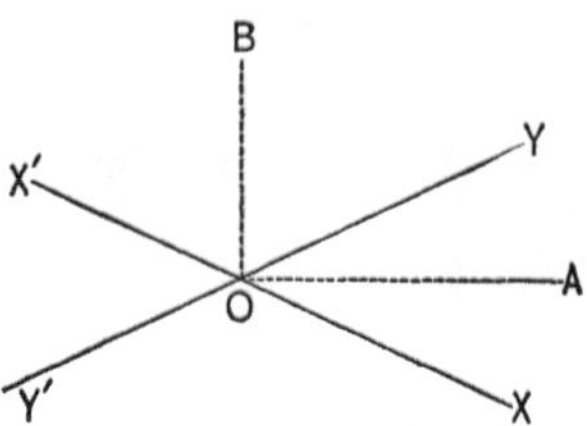

Given	two lines, XX', YY', cutting at O; also OA, OB, bisecting $\angle\!\!\!\!s\ XOY$, YOX', respectively.

To prove	that $OA \perp OB$.

Proof. 1.	$\angle AOY = \tfrac{1}{2} \angle XOY$.	Given; § 18

2.	$\angle YOB = \tfrac{1}{2} \angle YOX'$.	Given; § 18

3.	$\therefore \angle AOB = \tfrac{1}{2} \angle XOX'$.	Ax. 2

(If equals are added to equals, the sums are equal.)

4.	$\therefore \angle AOB = \tfrac{1}{2}$ of a st. $\angle$.	Def. st. $\angle$

(§ 14. If the two arms of an $\angle$ lie in the same st. line on opposite sides of the vertex, a st. $\angle$ is said to be formed.)

5.	$\therefore \angle AOB =$ a rt. $\angle$.	Def. rt. $\angle$

(§ 19. A rt. $\angle$ is half of a st. $\angle$.)

6.	$\therefore OA \perp OB$.	Def. $\perp$

(§ 20. If two lines meet and form a rt. $\angle$, each line is said to be $\perp$ to the other.)

Proposition X.

44. Theorem. *The bisectors of the four angles which two intersecting lines make with each other form two straight lines.*

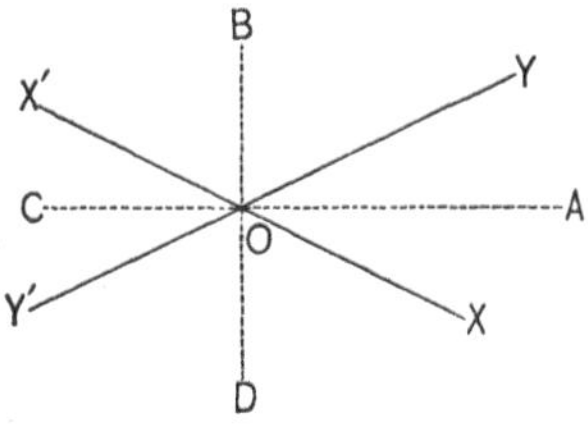

Given XX' intersecting YY' at O, OA bisecting $\angle XOY$, OB bisecting $\angle YOX'$, OC bisecting $\angle X'OY'$ and OD bisecting $\angle Y'OX$.

To prove that COA and DOB are straight lines.

Proof. 1. $\angle AOB$ and BOC are rt. $\angle$s.   Prop. IX
                (State prop. IX.)

2. ∴ the two together form a st. angle.   Def. rt. $\angle$
                (§ 19. State the definition.)

3. ∴ COA is a st. line.   Def. st. $\angle$
                (§ 14. State the definition.)

4. Similarly for DOB.

---

45. The nature of a logical proof should now be understood. Before continuing, however, the following points should be emphasized:

a. Every statement in a proof must be based upon a postulate, an axiom, a definition, or some proposition previously considered of which the student is prepared to give the proof again when he refers to it.

b. No statement is true simply because it appears to be true from a figure which the student may have drawn, no matter how carefully. Many cases will be found, for example, where angles appear equal when they are not so.

c. The arrangement of the discussion of a theorem is as follows:

Given. Here is stated, with reference to the figure which accompanies the proof, whatever is given by the theorem.

To prove. Here is stated the exact conclusion to be derived from what is given.

Proof. Here are set forth, in concise steps, the statements to prove the conclusion just asserted. If the proof is written on the blackboard, the steps should be numbered for convenient reference by class and teacher. The teacher will state how much in the way of written or indicated authorities shall be required after each step.

Corollary. A **corollary** is a proposition so connected with another as not to require separate treatment. The proof is usually simple, but it must be given with the same accuracy as that of the proposition to which it is attached. It is usually sufficient to say, This is proved in step 4; or, This follows from steps 2 and 5 by axiom 3, etc. In every case the student should (1) clearly prove the corollary, but (2) do so as concisely as possible. A corollary may also follow from a definition; thus, from the definitions of *Proposition* and *Theorem* the following might be stated as a corollary: Every theorem is a proposition, but not every proposition is a theorem; and as a part of our definition of a *Perigon* we incorporated the corollary (the term then being undefined) that a perigon equals two straight angles.

Note. Any item of interest may be inserted under this head.

Exercises. **13.** Of the proofs of the preliminary theorems, state which are direct and which indirect. (See note on p. 14.)

14. How can you form a right angle by paper folding? Prove it.

BOOK I. — RECTILINEAR FIGURES.

1. TRIANGLES.

46. A **figure** is any combination of lines and points formed under given conditions.

E.g. an angle is a figure, for it is a combination of two lines and one point formed under the condition that the two lines proceed from the point.

47. A **rectilinear figure** is a figure of which all the lines are straight.

Plane geometry treats of figures in one plane, — *plane figures.*

Hence in plane geometry, which in this work extends through Books I to V inclusive, the word *figure* used alone denotes a *plane figure*, and all propositions and definitions refer to such figures placed in one plane.

48. If the two end-points of a broken line coincide, the figure obtained is called a **polygon,** and the broken line its **perimeter.** The vertices of the angles made by the segments of the perimeter are called the **vertices of the polygon,** and the segments between the vertices are called the **sides** of the polygon.

49. The perimeter of a polygon divides the plane into two parts, one finite (the part inclosed), the other infinite. The finite part is called *the surface of the polygon,* or for brevity simply *the polygon.*

A point is said to be *within* or *without* the polygon according as it lies within or without this finite part.

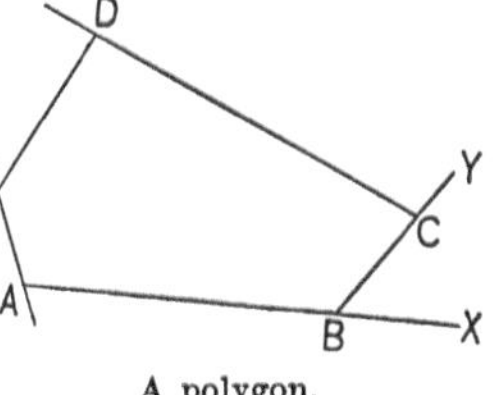

A polygon.

The figure *ABCDE* is a polygon (the sides being produced for a subsequent definition).

50. In passing counter-clockwise around the perimeter of a polygon the angles on the left are called the **interior angles** of the polygon, or for brevity simply *the angles of the polygon.*

Such are the angles CBA, DCB, EDC, ... in the figure on p. 21.

51. If the sides of a polygon are produced in the same order, the angles between the sides produced and the following sides are called the **exterior angles** of the polygon.

Such are the angles XBC, YCD, ... in the figure on p. 21. They are the angles through which one would turn, at the successive corners, in walking around the polygon.

52. A line joining the vertices of any two angles of a polygon which have not a common arm, is called a **diagonal.**

Such a line would be the one joining A and C in the figure on p. 21. The sides, angles, and diagonals of a polygon are often called its *parts.*

53. A polygon which has all of its *sides* equal is called **equilateral.**

A polygon which has all of its *angles* equal is called **equiangular.**

54. Two polygons are said to be **mutually equilateral,** or one is said to be *equilateral* to the other, when the *sides* of the one are respectively equal to the *sides* of the other.

Two polygons are said to be **mutually equiangular,** or one is said to be *equiangular* to the other, when the *angles* of the one are respectively equal to the *angles* of the other.

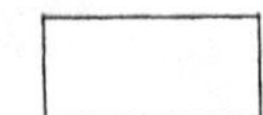 

55. A polygon of three sides is called a **triangle**; one of four sides, a **quadrilateral.**

56. Any side of a polygon may be called its **base,** the side on which the figure appears to stand being usually so called, as AB in the figure on p. 21.

In the case of a triangle, the vertex of the angle opposite the base is called the **vertex of the triangle,** the angle itself being called the *vertical angle of the triangle,* and the other two angles the *base angles.*

Thus, in the first triangle on p. 25, C is the vertex of the triangle, $\angle C$ is the vertical angle, $\angle A$ and $\angle B$ are the base angles.

57. Two figures which may be made to coincide in all their parts by being placed one upon the other are said to be **congruent.**

For example, two line-segments may be congruent, or two angles, or two triangles, etc.

58. The operation of placing one figure upon the other so that the two shall coincide is called *superposition,* and the figures are sometimes called *superposable* (a synonym of congruent).

This is illustrated in prop. I.

Superposition is an imaginary operation. It is assumed as a postulate (§ 61) that figures may be moved about in space with no other change than that of position. The actual movement is, however, left for the imagination.

59. It will hereafter be explained and defined that polygons of the same shape are called *similar,* the symbol of similarity being ∽, and that those of the same area are called *equal* or *equivalent,* the symbol being $=$. *Congruent figures are both similar and equal,* and hence the symbol for congruence is ≅, a symbol used in modified form by the great mathematician Leibnitz.

The symbol ∽ is derived from the letter S, the initial of the Latin *similis,* similar.

Many writers use *equal* for *congruent,* and *equivalent* for *equal,* as above defined. But because of the various meanings of the word *equal,* and its general use as a synonym for

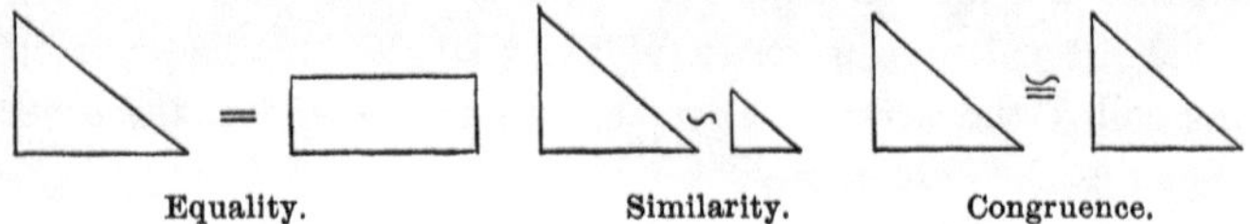

Equality. Similarity. Congruence.

equivalent, the more exact word *congruent* with its suggestive symbol is coming to be employed. The student should be familiar with this other use of the words *equal* and *equivalent.*

60. It is customary to designate the sides of a triangle by the small letters corresponding to the capital letters which designate the opposite vertices.

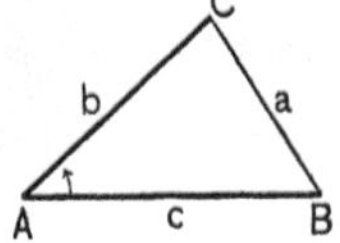

Thus, in the figure, side a is opposite vertex A, etc.

61. It now becomes necessary to assume three other postulates.

Postulates of Motion.

1. *A figure may be moved about in space with no other change than that of position, and so that any one of its points may be made to coincide with any assigned point in space.*

That is, we may pick up one polygon and place it on another without changing its shape or size.

2. *A figure may be moved about in space while one of its points remains fixed.*

Such movement is called *rotation about a center,* the center being the fixed point.

3. *A figure may be moved about in space while two of its points remain fixed.*

Such movement is called *rotation about an axis,* the axis being the line determined by the two points.

Proposition I.

62. Theorem. *If two triangles have two sides and the included angle of the one respectively equal to two sides and the included angle of the other, the triangles are congruent.*

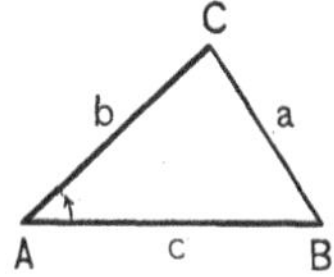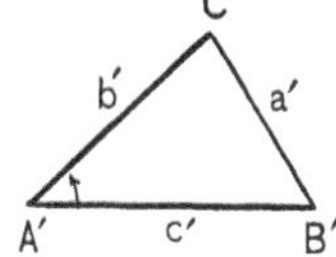

Given   the $\triangle$ ABC, $A'B'C'$ such that

$$c = c',$$
$$b = b', \text{ and}$$
$$\angle A = \angle A'.$$

To prove that $\triangle ABC \cong \triangle A'B'C'$.

Proof. 1. Place $\triangle A'B'C'$ on $\triangle ABC$ so that

A' falls on A, and	§ 61, 1
c' coincides with its equal c.	§ 61, 2

2. Then b' may be caused to fall on b,

because $\angle A' = \angle A$.	Given; § 61, 3

3. Then C' will fall at C,

because $b' = b$.	Given; § 57

4. $\therefore a'$ will coincide with a.   § 28, 1

(Two points determine a straight line.)

5. $\therefore \triangle ABC \cong \triangle A'B'C'$, by definition of congruence.

§ 57

Notes. This is a proof by superposition.

The theorem may be stated, A triangle is *determined* when two sides and the included angle are given.

In the exercises hereafter given, the proofs are to be given in full; when a question is asked, a proof of the answer is to be given; when a theorem is suggested, it is to be completely stated and then proved.

Proposition II.

63. Theorem. *If two triangles have two angles and the included side of the one respectively equal to two angles and the included side of the other, the triangles are congruent.*

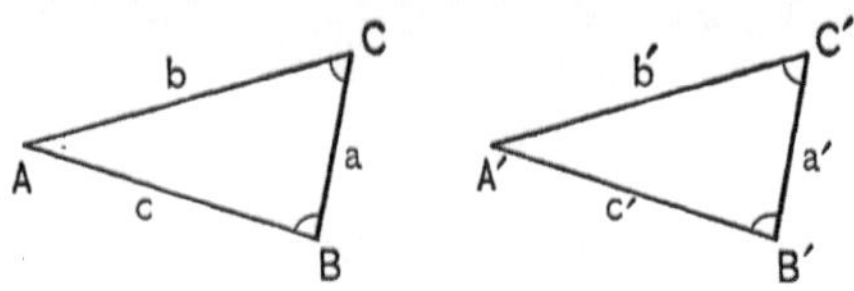

Given   the $\triangle ABC$ and $A'B'C'$ such that

$$\angle C = \angle C',$$
$$\angle B = \angle B', \text{ and}$$
$$a = a'.$$

To prove   that $\triangle ABC \cong \triangle A'B'C'.$

Proof. 1. Place $\triangle A'B'C'$ on $\triangle ABC$ so that a' falls on a and $\angle C'$ coincides with its equal $\angle C$.   § 61

2. Then B' will fall on B because $a' = a$.   Given

3. Then c' will fall on c because $\angle B' = \angle B$.   Given

4. $\therefore A'$ will coincide with A.   § 28, 2

    (Two straight lines determine a point.)

5. $\therefore \triangle ABC \cong \triangle A'B'C'$, by definition of congruence.
     § 57

Note.   Prop. II, and prop. III following, are attributed to Thales.

Exercises. 15. In the figure on p. 19, given that OA bisects angle XOY, and that OB is perpendicular to OA, prove that OB bisects angle YOX'.

16. Show that the distance BA across a lake may be measured by setting up a stake at O, sighting across it to fix the lines $A'B$ and $B'A$, laying off $OA' = OA$, and $OB' = OB$, and then measuring $B'A'$.

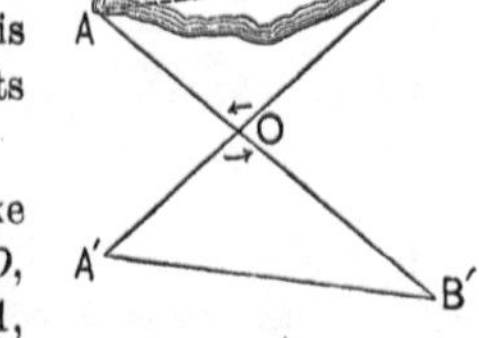

64. **Reciprocal Theorems.** The student will notice that propositions I and II have a certain similarity. Indeed, if the words *side* and *angle* are interchanged in prop. I, it becomes prop. II, and if interchanged in prop. II that becomes prop. I. Theorems of this kind are called **reciprocal.** The relation is more clearly seen by resorting to parallel columns.

Prop. I. If two triangles have two *sides* and the included *angle* of the one respectively equal to two *sides* and the included *angle* of the other, the triangles are congruent.

Prop. II. If two triangles have two *angles* and the included *side* of the one respectively equal to two *angles* and the included *side* of the other, the triangles are congruent.

Moreover, if small letters and capitals are interchanged ·in the proof of prop. I, the proof becomes that of prop. II.

65. The principle involved is called the **Principle of Reciprocity,** and is extensively used in geometry. But the student must not suppose that because a theorem is true its reciprocal theorem is also true; in elementary geometry, involving measurements, the reciprocal is often false. The principle is, however, of great value even here, for it leads the student to see the relation between propositions, and *it often suggests new possible theorems for investigation.* For these purposes we shall use it.

At present it is sufficient to say that for many theorems of plane geometry reciprocal theorems may be formed by replacing the words

point by line,

line by point,

angles of a triangle by (opposite) sides of a triangle,

sides of a triangle by (opposite) angles of a triangle.

Exercises. **17.** Explain this statement and tell why it is true : Any two sides and the included angle of a triangle determine the remaining parts.

18. State the reciprocal of ex. 17 and tell whether it is true, and why.

Proposition III.

66. **Theorem.** *If two sides of a triangle are equal, the angles opposite those sides are equal.*

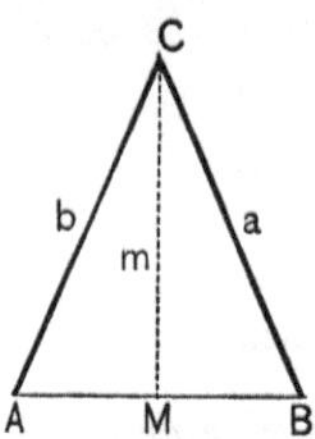

Given the $\triangle ABC$ with $AC = BC$.

To prove that $\angle A = \angle B$.

Proof. 1. Suppose m to bisect $\angle ba$.

2. Then $\because b = a$, Given

  and $\angle bm = \angle ma$,

  and $m \equiv m$,

3. $\therefore \triangle AMC \cong \triangle BMC$, Prop. I

  (State prop. I.)

  and $\angle A = \angle B$, by definition of congruence. § 57

Corollary. *If a triangle is equilateral, it is also equiangular.*

For by the theorem the angles opposite the equal sides are equal.

67. **Definitions.** The line from any vertex of a triangle to the mid-point of the opposite side is called the **median** to that side.

In the above figure, CM is the median to AB.

If a triangle has two equal sides, it is called an **isosceles triangle.**

The third side is called the *base of the isosceles triangle,* and the equal sides are called the *sides.*

A triangle which has no two sides equal is called a **scalene triangle.**

The **distance** from one point to another is the length of the straight line-segment joining them.

The **distance** from a point to a line is the length of the perpendicular from that point to that line.

That this perpendicular is unique will be proved later.

This is the meaning of the word *distance* in plane geometry. In speaking of points on a curved surface (for example, the earth's surface), distance may be measured on a curved line.

68. In the figure of prop. III,

$$\triangle\, AMC \cong \triangle\, BMC, \text{ as proved.}$$

$$\therefore AM = MB,$$

$$\text{and } \angle\, CMA = \angle\, BMC,$$

and hence each is a right angle.

In cases of this kind the points A and B are said to be **symmetric with respect to an axis.** Hence, in the figure, CM is called an **axis of symmetry.** And, in general, two *systems of points,* A_1, B_1, C_1, ..., A_2, B_2, C_2, ..., are said to be *symmetric with respect to an axis* when all lines, A_1A_2, B_1B_2, ..., are bisected at right angles by that axis.

Also, *two figures are said to be symmetric with respect to an axis* when their systems of points are symmetric.

A *single figure,* like that of prop. III, is said to be *symmetric with respect to an axis* when this axis divides it into two symmetric figures.

Exercises. **19.** If four lines go out from a point making four angles of which the first and third are equal, and the second and fourth are equal, prove that the four lines form two intersecting straight lines.

20. In the figure on p. 19, if a line passes through O and bisects angle XOA, prove that it also bisects angle $X'OC$.

Proposition IV.

69. Theorem. *If two angles of a triangle are equal, the sides opposite those angles are equal.*

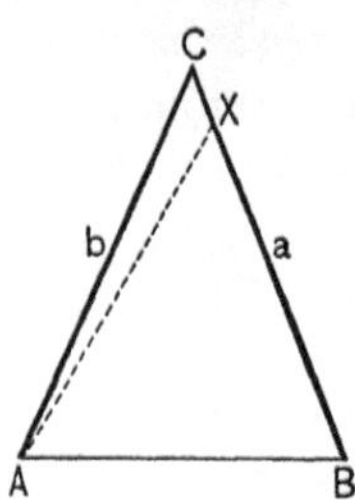

Given   the $\triangle ABC$ with $\angle A = \angle B$.

To prove that $a = b$.

Proof. 1. Suppose that $a \neq b$,

and that $a > b$.

2. Then let BX, a part of a, equal b, and join A and X.

3. Then $\because \angle B = \angle BAC$,   Given

and $AB \equiv AB$,

$\therefore \triangle ABC \cong \triangle BAX$.   Why ?

4. $\therefore$ the supposition leads to an absurdity, for

$$\triangle ABC > \triangle BAX, \qquad \text{Ax. 8}$$

(State ax. 8.)

and $\therefore a \not> b$.

In the same way it may be shown that $a \not< b$,

and $\therefore a = b$.

Corollary. *If a triangle is equiangular, it is also equilateral.* (Why ?)

Exercise. 21. If four points, A, B, C, D, are placed in order on a line, and if $AC = BD$, prove that $AB = CD$.

Proposition V.

70. Theorem. *If any side of a triangle is produced, the exterior angle is greater than either of the interior angles not adjacent to it.*

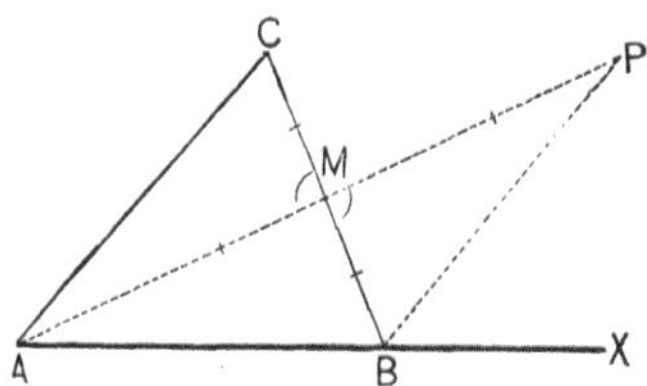

Given   the $\triangle ABC$, with AB produced to X.

To prove   that $\angle XBC > \angle C$, and also $> \angle BAC$.

Proof. 1. Suppose BC bisected at M, AM drawn and produced to P so that $MP = AM$, and BP drawn.

2. Then $\because \angle BMP = \angle CMA$,   Why ?

   $\therefore \triangle BPM \cong \triangle CAM$,   Why ?

   and $\angle PBM = \angle C$.   § 57

3. But $\angle XBC > \angle PBM$.   Why ?

4. $\therefore \angle XBC > \angle C$.   Why ?

5. Similarly, by producing CB, bisecting AB at N, producing CN, etc., it can be shown that an angle equal to $\angle XBC$ is greater than $\angle BAC$.

Exercises.   22. Show that, in the figure of prop. V, $\angle XBC > \angle BAC$ by following out in full the proof suggested in step 5.

23. In the figure of prop. V, join C to any point in the segment AB and prove that $\angle CBA + \angle BAC < 180°$.

24. If a diagonal of a quadrilateral bisects two angles, the quadrilateral has two pairs of equal sides.

25. How many equal lines can be drawn from a given point to a given line ?   Show that if another is supposed to be drawn, an absurdity results.

Proposition VI.

71. Theorem. *If two sides of a triangle are unequal, the opposite angles are unequal and the greater side has the greater angle opposite.*

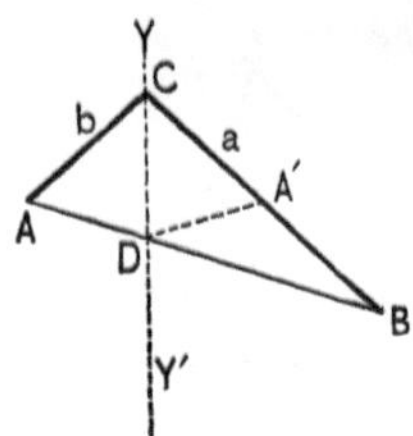

Given   the $\triangle ABC$, with $a > b$.

To prove that $\angle A > \angle B$.

Proof. 1. Suppose $\angle C$ bisected by YY' cutting AB at D, CA' made equal to CA, and DA' drawn.

     2. Then $\triangle ADC \cong \triangle A'DC$, and $\angle A = \angle CA'D$. Why?

     3. But $\angle CA'D > \angle B$.   Prop. V

(§ 70. If any side of a $\triangle$ is produced, the exterior angle is greater than either of the int. $\angle$s not adjacent to it.)

     4. $\therefore \angle A > \angle B$.   Subst. 2 in 3

Exercises. **26.** State, without proof, the reciprocal of prop. VI.

27. Can a scalene triangle have two equal angles? Proof.

28. Prove prop. VI by drawing AA' instead of DA', and proving that $\angle A > \angle A'AC = \angle CA'A > \angle B$.

29. $ABCD$ is a quadrilateral of which DA is the longest side and BC the shortest. Which is greater, $\angle B$ or $\angle D$? Prove it. (Suggestion: Draw BD.) Also $\angle C$ or $\angle A$? Prove it.

30. How many perpendiculars can be drawn to a given line from a point outside that line? Show that any other supposition violates prop. V.

31. ABC is a triangle having $\angle B =$ twice $\angle A$; $\angle B$ is bisected by a line meeting b at D; prove that $AD = BD$.

Proposition VII.

72. Theorem. *If two angles of a triangle are unequal, the opposite sides are unequal and the greater angle has the greater side opposite.*

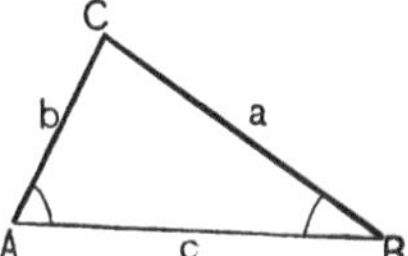

Given   the $\triangle ABC$ with $\angle A > \angle B$.

To prove that $a > b$.

Proof. 1. $a \neq b$, for if $a = b$, then $\angle A = \angle B$.   Why ?

2. $a \nless b$, for if $a < b$, then $\angle A < \angle B$.

Prop. VI.   State it.

3. $\therefore a$ must be greater than b.

Note. It must not be inferred from props. VI, VII that, because one angle of a triangle is twice as large as another, one side is twice as long as another.

Exercises. 32. Prove that if the bisector of any angle of a triangle is perpendicular to the opposite side, the triangle is isosceles.

33. Suppose any point taken on the perpendicular bisector of a line; is it equally or unequally distant from the ends of the line ? Give the proof in full.

34 *a*. Prove that in an isosceles triangle ABC, where $a = b$, the bisector of $\angle C$, produced to c, bisects *side c*.

34 *b*. Prove that in an isosceles triangle abc, where $\angle A = \angle B$, the bisector of *side c*, joined to C, bisects $\angle C$.

35. After reading § 73, state the converse of each of the following : (a) prop. III ; (b) prop. IV ; (c) prop. VI ; (d) prop. VII ; (e) this statement, If the animal is a horse, then the animal has two eyes. Of these converses, how many are true ?

36. What kind of a triangle is formed by joining the mid-points of the sides of an equilateral triangle ? Prove it.

73. The Law of Converse. Two theorems are said to be the **converse,** each of the other, when what is given in the one is what is to be proved in the other, and *vice versa.*

E.g. props. VI and VII. The converse of a theorem must not be confused with its reciprocal. Props. I and II are reciprocal, but not converse.

Because a theorem is true its converse is not necessarily true.

For example, prel. prop. I may be stated thus : Given that $\angle r$ and r' are rt. $\angle$, to prove that $\angle r = \angle r'$; the converse is, Given that $\angle r = \angle r'$, to prove that they are rt. $\angle$. This converse is evidently false, for $\angle r$ could equal $\angle r'$ without their being rt. $\angle$.

But there is one important class of converse theorems, illustrated by props. IV and VII, that should be mentioned. *Whenever three theorems have the following relations, their converses must be true :*

1. If it has been proved that when $A > B$, then $X > Y$, and
2. " " " $A = B$, " $X = Y$, "
3. " " " $A < B$, " $X < Y$,

then the converse of each of these is true. For

1'. If $X > Y$, then A can neither be equal to nor less than B, without violating 2 or 3 ; $\therefore A > B$. (Converse of 1.)

2'. If $X = Y$, then A can neither be greater nor less than B, without violating 1 or 3 ; $\therefore A = B$. (Converse of 2.)

3'. If $X < Y$, then A can neither be greater than nor equal to B, without violating 1 or 2 ; $\therefore A < B$. (Converse of 3.)

The law just proved will hereafter be referred to as the **Law of Converse.** By its use the proof of the converse of many theorems, where true, is made very simple.

The student should not proceed further unless the Law of Converse is thoroughly understood, and its proof mastered.

Prop. VII may now be proved by the Law of Converse, thus :

$$\text{If } a > b, \text{ then } \angle A > \angle B. \qquad \text{Prop. VI}$$
$$\text{If } a = b, \text{ '' } \angle A = \angle B. \qquad \text{'' III}$$
$$\text{If } a < b, \text{ '' } \angle A < \angle B. \qquad \text{'' VI}$$

$\therefore$ each converse is true, and if $\angle A > \angle B$, then $a > b$.

74. Suggestions as to the Treatment of the Exercises. Thus far the student has been left to his own ingenuity in treating the exercises. A few suggestions should now be given.

1. *In attacking a theorem take the most general figure possible.*

E.g. if a theorem relates to a triangle, draw a *scalene* triangle ; an equilateral or an isosceles triangle often deceives the eye, and leads away from the demonstration. *Draw all figures accurately ;* an accurate figure often *suggests* the demonstration. But the student who relies too much upon the accuracy of the figure in the demonstration itself is liable to go astray.

2. *Be certain that what is given and what is to be proved are clearly stated, with reference to the letters of the figure.*

This has been done in all of the theorems thus far proved. The neglect to do so in the exercises is one of the most fruitful sources of failure.

3. *Then begin by assuming the theorem true ; see what follows from that assumption ; then see if this can be proved true without the assumption ; if so, try to reverse the process.*

E.g. suppose $PO \perp X'X$, and PB, PA two obliques cutting off $OA > OB$, as in the figure, and that it is required to prove $PA > PB$. Assume it true ; then $\angle b > \angle a$. Now see if $\angle b > \angle a$ *without* the assumption ; $\angle b > \angle c$, which $= \angle d$, which $> \angle a$, by prop. V ; $\therefore \angle b > \angle a$, without the assumption. Now reverse the process ; $\because \angle b > \angle a$, $\therefore PA > PB$ by prop. VII.

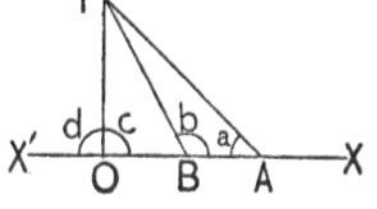

4. *Or begin by assuming the theorem false, and endeavor to show the absurdity of the assumption. (Reductio ad absurdum.)*

5. To secure a clearer understanding of the theorem it is often well to follow Pascal's advice and *substitute the definition for the name of the thing defined.*

E.g. suppose it is to be proved that the median to the base of an isosceles triangle is perpendicular to the base. Instead of saying :

" Given CM the median to the base of the isosceles triangle ABC " (see figure on p. 28), it is often better to say :

" Given $\triangle ABC$, with $AC = BC$, and M taken on AB so that $AM = MB$," for then the facts stand out prominently without any confusing terms.

Pʀᴏᴘᴏsɪᴛɪᴏɴ VIII.

75. Theorem. *The sum of any two sides of a triangle is greater than the third side.*

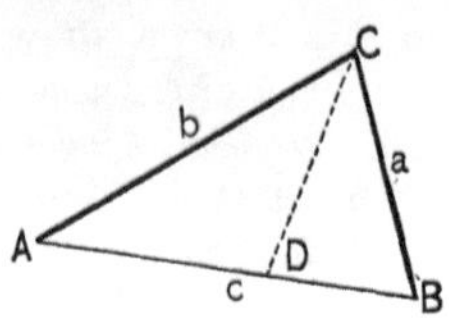

Given the $\triangle ABC$.

To prove that $a + b > c$.

Proof. 1. Suppose $\angle C$ bisected by CD.

Then $\angle CDA > \angle DCB$. Prop. V. State it

2. And $\quad \because \angle ACD = \angle DCB,$ Step 1

$\therefore \angle CDA > \angle ACD.$

$\therefore b > AD.$ Prop. VII. State it

Similarly, $a > DB.$

3. $\therefore a + b > c.$

Cᴏʀᴏʟʟᴀʀʏ. *The difference of any two sides of a triangle is less than the third side.*

For if $a + b > c$, and $c > b$, then $a > c - b$, by ax. 5.

Exercises. 37. Two equal lines, AC and AD, are drawn on opposite sides of a line AB and making equal angles with it ; BC and BD are drawn. Show that BC and BD also make equal angles with AB.

38. P, Q, R are points on the sides AB, BC, CA, respectively, of an equilateral triangle ABC, such that $AP = BQ = CR$; joining P, Q, and R, prove that $\triangle PQR$ is equilateral. (Notice that ex. 36 is merely a special case of this one.)

39 *a*. The bisectors of the equal *angles* of an isosceles triangle form, with the *base*, an isosceles triangle.

39 *b*. The mid-points of the equal *sides* of an isosceles triangle form, with the *vertex*, the vertices of an isosceles triangle.

Proposition IX.

76. Theorem. *If from the ends of a side of a triangle two lines are drawn to a point within the triangle, their sum is less than the sum of the other two sides of the triangle, but they contain a greater angle.*

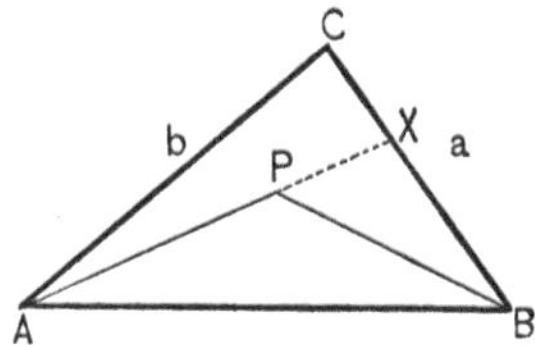

Given   the $\triangle ABC$, P a point within, and BP and PA drawn.

To prove   that (1) $BP + PA < a + b$, (2) $\angle APB > \angle C$.

Proof. 1. Produce AP to meet a at X.

Then
$$XP + PA = XA < XC + b, \quad \text{Ax. 8; prop. VIII}$$
(State ax. 8 and prop. VIII.)

and $\qquad\qquad BP < BX + XP.$ $\qquad\qquad$ Prop. VIII

2. $\therefore BP + XP + PA < BX + XC + XP + b.$

3. $\therefore BP \qquad\quad + PA < \qquad a \quad + \qquad b,$
which proves (1). Why?

4. Also,
$$\angle APB > \angle PXB > \angle C, \text{which proves (2). Why?}$$

Exercises. 40 a. If the equal *sides* of an isosceles triangle are bisected, the *lines* joining the *points* of bisection with the vertices of the equal *angles* are equal.

40 b. If the equal *angles* of an isosceles triangle are bisected, the *angles* formed by the *lines* of bisection and the equal *sides* are equal.

41. The perimeter of a quadrilateral is less than twice the sum of its two diagonals.

Proposition X.

77. Theorem. *If two triangles have two sides of the one respectively equal to two sides of the other, but the included angles unequal, then the third sides are unequal, the greater side being opposite the greater angle.*

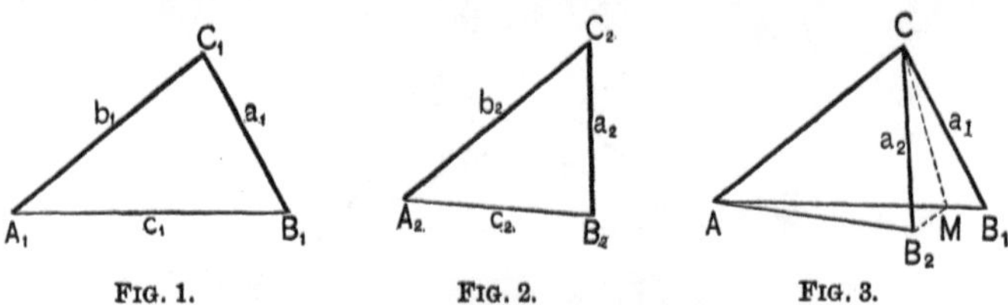

Given the $\triangle A_1B_1C_1$ and $A_2B_2C_2$, with $a_1 = a_2$, $b_1 = b_2$, but $\angle C_1 > \angle C_2$.

To prove that $c_1 > c_2$.

Proof. 1. Suppose $\triangle A_2B_2C_2$ placed on $\triangle A_1B_1C_1$ so that b_2 and b_1, being equal, coincide. § 61

Then $\because \angle C_1 > \angle C_2$, side a_2 must fall within $\angle C_1$, as in Fig. 3.

2. Suppose CM drawn bisecting $\angle B_2CB_1$, and B_2M drawn.

3. Then in $\triangle B_1MC, B_2MC$,

$$CB_1 = CB_2, \qquad\qquad \text{Given}$$

$$CM \equiv CM,$$

$$\angle MCB_1 = \angle B_2CM. \qquad\qquad \text{Step 2}$$

4. $\qquad\qquad \therefore \triangle B_1MC \cong \triangle B_2MC,$

and $\qquad\qquad MB_1 = MB_2.$ Prop. I

5. But $\quad AM + MB_2 > AB_2.$ Prop. VIII

$$\therefore AM + MB_1 > AB_2,$$

or $\qquad\qquad c_1 > c_2.$

The proof is the same when B_2 falls above A_1B_1.

Proposition XI.

78. Theorem. *If two triangles have two sides of the one respectively equal to two sides of the other, but the third sides unequal, then the included angles are unequal, the greater angle being opposite the greater third side.*

Given $\triangle$ $A_1B_1C_1$ and $A_2B_2C_2$, with $a_1 = a_2$, $b_1 = b_2$, $c_1 > c_2$.

To prove   that $\angle C_1 > \angle C_2$.

Proof.   1. It has been shown that if $a_1 = a_2$, $b_1 = b_2$,
          and if $\angle C_1 > \angle C_2$, then $c_1 > c_2$.   Prop. X

     2. And if $\angle C_1 =$ "   "   " $=$ "   Prop. I

     3.   "   " $\angle C_1 <$ "   "   " $<$ "   Prop. X

     4. $\therefore$ the converses are true, which proves the theorem.
                                 § 73. Law of Converse
(Explain the Law of Converse. Since this law is so often used, it should be reviewed frequently.)

Exercises.   **42.** Are props. X and XI reciprocals? converses?

43. In $\triangle ABC$, suppose $CA > AB$, and that points P, Q are taken on AB, CA respectively, so that $PB = CQ$. Prove that $BQ < CP$.

44. Investigate ex. 43 when P is taken on AB produced, and Q on AC produced.

45. The equal sides, AC, BC, of an isosceles triangle ABC are produced through the vertex to P and Q respectively, so that $AP = BQ$. Prove that $BP = AQ$.

46. Prove that the straight line joining any two points is less than any broken line joining them.

47. Prove that the perimeter of a triangle is less than twice the sum of the three medians.

48. In a quadrilateral, prove that the sum of either pair of opposite sides is less than the sum of its two diagonals.

49. If the perpendicular from any vertex of a triangle to the opposite side divides that side into two segments, how does each of these segments compare in length with its adjacent side of the triangle? Prove it.

Pʀᴏᴘᴏsɪᴛɪᴏɴ XII.

79. Theorem. *If two triangles have the three sides of the one respectively equal to the three sides of the other, the triangles are congruent.*

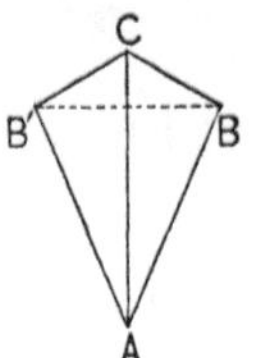

Given   △ *ABC*, *AB'C*, with *AB* = *AB'*, *BC* = *B'C*, and *AC* = *AC*.

To prove   that △ *ABC* ≅ △ *AB'C*.

Proof. 1. Suppose no side longer than *AC*. Then the △, being mutually equilateral, may be placed with *AC* in common, and on opposite sides of *AC*. Draw *BB'*.

2. Then   ∠ *CBB'* = ∠ *BB'C*,   Prop. III
   and   ∠ *B'BA* = ∠ *AB'B*.   Why ?

3.   ∴ ∠ *CBA* = ∠ *AB'C*.   Why ?

4.   ∴ △ *ABC* ≅ △ *AB'C*.   Why ?

   AC is evidently an axis of symmetry (§ 68) in the figure.

———

Exercises. 50. Suppose three sticks to be hinged together to form a triangle, could the sides be moved so as to change the angles? On what theorem does the answer depend? How would it be with a hinged quadrilateral?

51. Ascertain and prove whether or not a quadrilateral is determined when the four sides and either diagonal are given in fixed order.

52. Also when the four sides and one angle are given in fixed order.

53. How many braces would it take to stiffen a three-sided plane figure? four-sided? five-sided?

Proposition XIII.

80. Theorem. *If two triangles have two angles of the one respectively equal to two angles of the other, and the sides opposite one pair of equal angles equal, the triangles are congruent.*

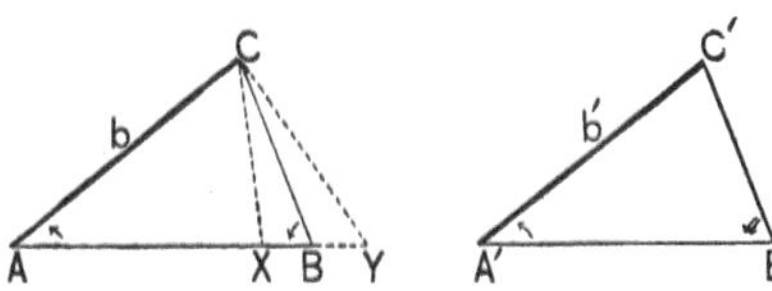

Given $\triangle ABC, A'B'C'$, with $\angle A = \angle A', \angle B = \angle B', b = b'$.

To prove that $\triangle ABC \cong \triangle A'B'C'$.

Proof. 1. Place $\triangle A'B'C'$ on $\triangle ABC$ so that A' falls at A, $A'B'$ lies along AB, and C and C' both lie on the same side of AB.   § 61

2. Then because $\angle A = \angle A'$, and $b = b'$, b' coincides with b, and C' with C.

3. Now B' cannot fall between A and B, as at X, for then $\angle CXA$, which $= \angle B'$, would be greater than $\angle B$.   Prop. V.   State it

4. Neither can B' fall on AB produced, as at Y, for then $\angle Y$, which $= \angle B'$, would be less than $\angle B$.   Prop. V

5. $\therefore B'$ must fall at B, and the $\triangle$ are congruent.   § 57

Exercises. **54.** If YO meets $X'X$ at O, and YA, YB are drawn meeting $X'X$ at A, B; and if $YA = YB$, and $AO \neq OB$, which is the greater, $\angle AYO$ or $\angle OYB$?

55. Consider the diagonals of an equilateral quadrilateral, (*a*) as to their bisecting each other, (*b*) as to the kind of angles they make with each other. State the theorems which you discover and prove them.

Pʀᴏᴘᴏꜱɪᴛɪᴏɴ XIV.

81. Theorem. *If two triangles have two sides of the one respectively equal to two sides of the other, and the angles opposite one pair of equal sides equal, then the angles opposite the other pair of equal sides are either equal or supplemental, and if equal the triangles are congruent.*

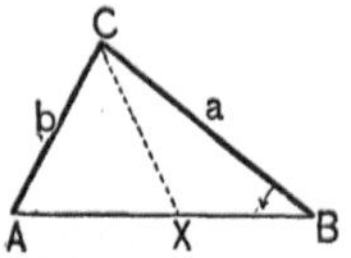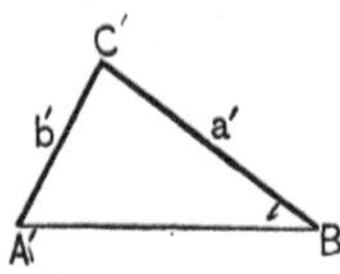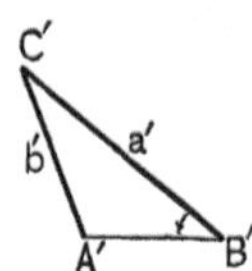

Given $\triangle ABC$, $A'B'C'$, with $a = a'$, $b = b'$, $\angle B = \angle B'$.

To prove that either (1) $\angle A = \angle A'$ and $\triangle ABC \cong \triangle A'B'C'$,

or   (2) $\angle A + \angle A' = $ st. $\angle$.

Proof. 1. Place $\triangle A'B'C'$ on $\triangle ABC$ so that B' falls at B, a' coincides with its equal a, and A' and A fall on the same side of a.   § 61

2. Then $\because \angle B = \angle B'$, $B'A'$ lies along BA.

3. Then either A' falls at A, the $\triangle$ are congruent and $\angle A = \angle A'$; or else A' falls at some other point on BA, as at X, and $\triangle A'B'C' \cong \triangle XBC$.

4. But $\because CX = b' = b$,

  $\therefore \angle A = \angle CXA$.   Prop. III. State it

5. And

  $\because \angle CXA + \angle BXC = $ st. $\angle$,   § 14, def. st. $\angle$

  $\therefore \angle A + \angle A' = $ st. $\angle$.

Exercises. **56.** In prop. XIV, step 3, X may lie on BA produced, in some cases. Draw the figure and prove.

57. In prop. XIV prove that $\angle A$ and $\angle A'$ must be equal, if (1) they are of the same species (*i.e.* both right, both acute, or both obtuse); or (2) angles B and B' are both right angles; or (3) $b \nless a$.

2. PARALLELS AND PARALLELOGRAMS.

82. Definitions.   If two straight lines in the same plane do not meet, however far produced, they are said to be **parallel**.

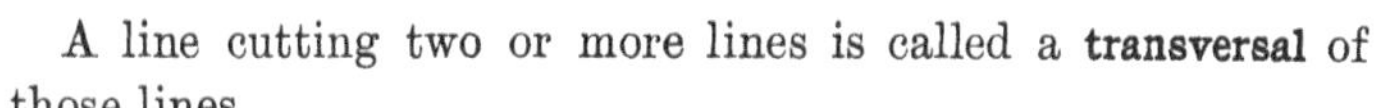

E.g. A and B in the annexed figure. The fact that A is parallel to B is indicated by the symbol $A \parallel B$.

A line cutting two or more lines is called a **transversal** of those lines.

In the figure of the parallel lines, T is a transversal of A and B. The adjacent figure shows a transversal of two non-parallel lines. The figure on p. 46 shows a transversal of three lines.

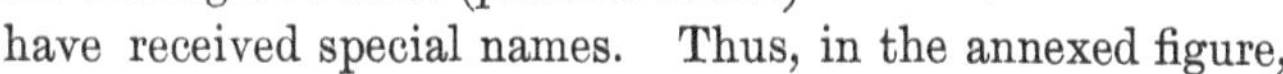

The angles formed by a transversal cutting two lines (parallel or not) have received special names.   Thus, in the annexed figure,

a, b, c', d' are called **exterior** angles ;

a', b', c, d are called **interior** angles ;

a and c' are called **alternate** angles ; also b and d', c and a', b' and d ;

a and a' are called **corresponding** angles ; also b and b', c and c', d and d'.

Exercises.   58. Angle A, of triangle ABC, is bisected by a line meeting BC at P. Which is the longer, AB or BP? Prove it. Also CA or PC? Prove it.

59. State the reciprocal of prop. XIII, and tell whether it is true without modification. In what proposition is your statement proved ?

60. If a quadrilateral has two pairs of equal sides, prove that it must have one pair and may have two pairs of equal angles, depending upon the arrangement of the sides.

Proposition XV.

83. Theorem. *If a transversal of two lines makes a pair of alternate angles equal, then (1) any angle is equal to its alternate angle, (2) any angle is equal to its corresponding angle, and (3) any two interior, or any two exterior, angles on the same side of the transversal are supplemental.*

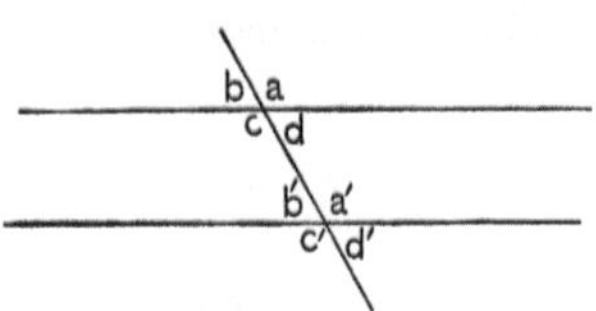

Given   a transversal cutting two lines, making equal alternate $\angle s$ d and b' as in the figure.

To prove that   (1) $\angle a = \angle c'$,

                 (2) $\angle a = \angle a'$,

                 (3) $\angle b + \angle c' = $ st. $\angle$.

Proof. 1. $\angle d + \angle a = $ st. $\angle$,

      and $\angle b' + \angle c' = $ st. $\angle$.   § 14, def. st. $\angle$

    2. $\therefore \angle d + \angle a = \angle b' + \angle c'$.   Why ?

    3. $\therefore \angle a = \angle c'$, which proves (1).   Ax. (?)

    4. $\because \angle c' = \angle a'$, $\therefore \angle a = \angle a'$, which proves (2).

    5. Now $\because \angle b' = \angle d$, and $\angle d = \angle b$,   Why ?

              $\therefore \angle b' = \angle b$.   Why ?

      But $\because \angle b' + \angle c' = $ st. $\angle$,   § 14, def. st. $\angle$

             $\therefore \angle b + \angle c' = $ st. $\angle$.

Corollaries.   1. *If two corresponding angles are equal, the same three conclusions follow.*

2. *If two interior or two exterior angles on the same side of the transversal are supplemental, the same conclusions follow.*

Proposition XVI.

84. Theorem.   *If a transversal of two lines makes a pair of alternate angles equal, the two lines are parallel.*

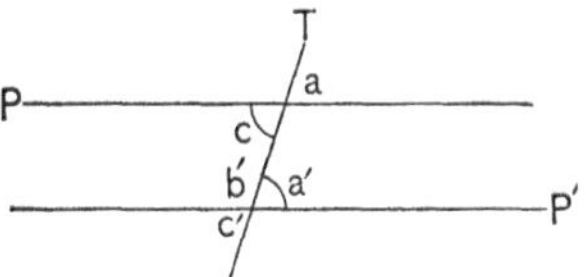

Given P and P', two lines, cut by a transversal T, making equal alternate $\measuredangle$ c and a'.

To prove   that $P \parallel P'$.

Proof.   1. P, P' cannot meet towards P', for then $\angle c$ would be an ext. $\angle$ of a $\triangle$, and $\therefore \angle c$ would be greater than $\angle a'$.   Prop. V.   State it

    2. P, P' cannot meet towards P, for then $\angle a'$ would be greater than $\angle c$.   Why ?

    3. $\therefore P$, P' cannot meet at all, and $P \parallel P'$.   Def. parallel
Similarly any other two alt. $\measuredangle$ may be taken equal.

CoROLLARIES.   1. *If two corresponding angles are equal, the lines are parallel.*

For then two alt. $\measuredangle$ are equal.   Prop. XV, cor. 1, which says — (?)

2. *If two interior or two exterior angles on the same side of the transversal are supplemental, the lines are parallel.*

For then two alt. $\measuredangle$ are equal.   Prop. XV, cor. 2, which says — (?)

3. *Two lines perpendicular to the same line are parallel.*
(Why ?)

___

Exercises.   61. In prop. XVI would lines bisecting $\angle a'$ and $\angle c$ be parallel ?   Prove it.

62. Show that if a draughtsman's square slides along a ruler, as in the annexed figure, $B_1C_1 \parallel B_2C_2$, and $A_1C_1 \parallel A_2C_2$.

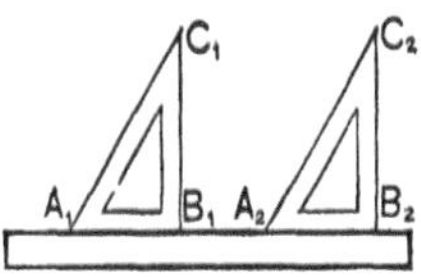

85. Postulate of Parallels. It now becomes necessary to assume another postulate, and upon it rests much of the elementary theory of parallels. It is: *Two intersecting straight lines cannot both be parallel to the same straight line.*

Cᴏʀᴏʟʟᴀʀʏ. *A line cutting one of two parallel lines cuts the other also, the lines being unlimited.*

(Show that the corollary is necessarily true if the postulate is.)

***

Pʀᴏᴘᴏsɪᴛɪᴏɴ XVII.

86. Theorem. *The alternate angles formed by a transversal with two parallels are equal.*

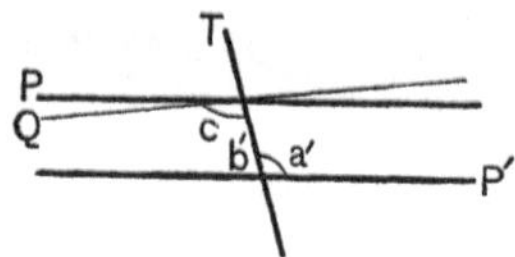

Given P and P' two parallels, and T, a transversal.

To prove that any $\angle c$ equals its alternate $\angle a'$.

Proof. 1. Suppose $\angle c > \angle a'$, and that Q is drawn as in the figure, making an $\angle$ equal to $\angle a'$.

    2. Then Q would be parallel to P'.   Why?

    3. But this would be impossible, $\because P \parallel P'$.   § 85

(Two intersecting straight lines cannot both be parallel to the same straight line.)

    4. Similarly, it is absurd to suppose that $\angle a' > \angle c$.

$$\therefore \angle c = \angle a'.$$

Cᴏʀᴏʟʟᴀʀɪᴇs. 1. *A line perpendicular to one of two parallels is perpendicular to the other also.*

For it cuts the other (§ 85, cor.) and the alternate angles are equal right angles.

2. *A line cutting two parallels makes corresponding angles equal, and the interior, or the exterior, angles on the same side of the transversal supplemental.*

For the alternate angles are equal (prop. XVII), and hence prop. XV applies.

3. *If the alternate or the corresponding angles are unequal, or if the interior angles on the same side of the transversal are not supplemental, then the lines are not parallel, but meet on that side of the transversal on which the sum of the interior angles is less than a straight angle.*

For the lines cannot be parallel, by prop. XVII and cor. 2.

Further, suppose $\angle c + \angle b' <$ st. $\angle$;
then $\quad \because \angle a' + \angle b' =$ st. $\angle$,
it follows that $\quad \angle c < \angle a'.$

$\therefore$ P and P' cannot meet towards P', for then $\angle c$ would be greater than $\angle a'$, prop. V.

Let the student give the proof in full form, in steps.

4. *Two lines respectively perpendicular to two intersecting lines cannot be parallel.*

For, in the annexed figure, let $AB \perp X$, $CD \perp Y$; join A and C. Then $\angle BAC <$ rt. $\angle$, and $\angle ACD <$ rt. $\angle$; $\therefore$ their sum is $<$ st. $\angle$; $\therefore$ cor. 3 applies. 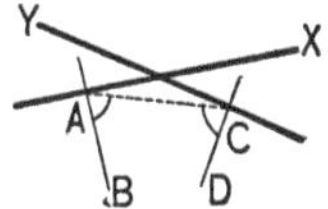

Give proof in full form in steps.

5. *If the arms of one angle are parallel or perpendicular to the arms of another, the angles are equal or supplemental.*

The proof is left to the student.

Exercises.   63. In the figure of prop. XV, suppose $a = c' = 120°\ 30'$, how large is each of the other angles ?

64. In the same figure, suppose $a + d' =$ st. $\angle$, and $a = 2\,d$, how large is each of the other angles ?

65. If a transversal cuts two lines making the sum of the two interior angles on the same side of the transversal a straight angle, one of them being $30°\ 27'$, how large is each of the other angles ?

Pʀᴏᴘᴏsɪᴛɪᴏɴ XVIII.

87. Theorem. *Lines parallel to the same line are parallel to each other.*

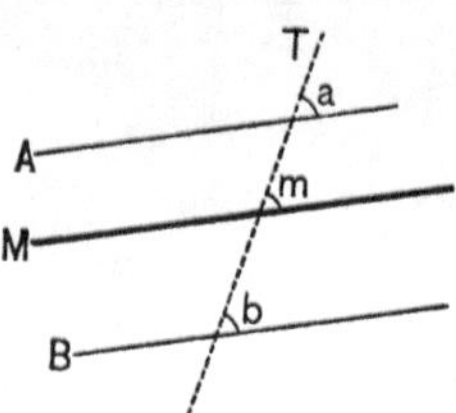

Given $A \parallel M$, and $B \parallel M$.

To prove that $A \parallel B$.

Proof. 1. Suppose T a transversal, making corresponding $\measuredangle\, a, m, b$, with A, M, B, respectively.

    2. Then $\because A \parallel M,$

             $\therefore \angle a = \angle m.$ Prop. XVII, cor. 2

    3. And $\because B \parallel M, \therefore \angle b = \angle m.$

    4. $\therefore \angle a = \angle b.$ Why ?

    5. $\therefore A \parallel B.$ Prop. XVI, cor. 1. State it

------

Exercises. 66. In prop. XVIII, if T cuts A, must it necessarily cut M? Why ?   If it cuts M, must it necessarily cut B?   Why ?

67. Prove that a line parallel to the base of an isosceles triangle makes equal angles with the sides or the sides produced.   (The line may pass above, through, or below the triangle, or through the vertex.)

68. If through any point equidistant from two parallels, two transversals are drawn, prove that they will cut off equal segments of the parallels.

69. ABC is a triangle, and through P, the point of intersection of the bisectors of $\angle B$ and $\angle C$, a line is drawn parallel to BC, meeting AB at M, and CA at N.   Prove that $MN = MB + CN$.

70. Through the mid-point of the segment of a transversal cut off by two parallels, a straight line passes, terminated by the parallels.   Prove that this line is bisected by the transversal.

Proposition XIX.

88. Theorem. *In any triangle, (1) any exterior angle equals the sum of the two interior non-adjacent angles ; (2) the sum of the three interior angles is a straight angle.*

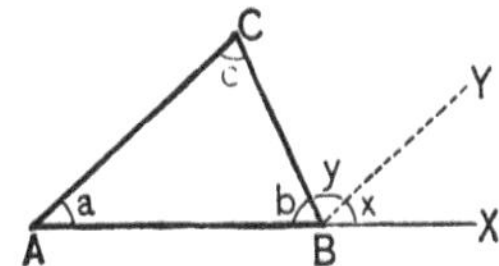

Given $\triangle ABC$, with AB produced to X.

To prove that (1) $\angle XBC = \angle A + \angle C$;

(2) $\angle A + \angle B + \angle C =$ st. $\angle$.

Proof. 1. Suppose $BY \parallel AC$, and $\angle\!\!\angle$ named as in the figure.

2. Then $\angle x = \angle a$, Why ?

and $\angle y = \angle c$. Why ?

3. $\therefore \angle x + \angle y$, or $\angle XBC$, $= \angle a + \angle c$,
which proves (1). Ax. 2

4. But $\angle x + \angle y + \angle b =$ st. $\angle$. Def. st. $\angle$

5. $\therefore \angle a + \angle b + \angle c =$ st. $\angle$,
by substituting 3 in 4, which proves (2).

Notes. 1. Prop. XIX, (2), is attributed to Pythagoras.

2. The theorem is one of the most important of geometry. To it and to its corollaries (p. 50) frequent reference is hereafter made.

Exercises. 71. PQR is a triangle having $PQ = PR$; RP is produced to S so that $PS = RP$; QS is drawn. Prove that $QS \perp RQ$.

72. Prove prop. XIX, (2), by drawing through C, in the figure given, a line $\parallel AB$.

73. Also by assuming any point P on AB, drawing PC, and showing that $\angle BPC + \angle CPA =$ st. $\angle$, and also equals the sum of the interior angles.

74. State the reciprocal of prop. VIII, and prove or disprove it.

Corollaries to prop. XIX.   1. *If a triangle has one right angle, or one obtuse angle, the other angles are acute.*

For the sum of all three is a straight angle.

2. *Every triangle has at least two acute angles.*

For if it had none or only one, the sum of the others would equal or exceed what kind of an angle, and thus violate what theorem ?

3. *From a point outside a given line not more than one perpendicular can be drawn to that line.*

For if two could be drawn, a triangle could be formed having how many right angles, thus violating what corollary ?

4. *If a triangle has a right angle, the two acute angles are complemental.*

For the sum of all three must equal two right angles; therefore, etc.

5. *If two triangles have two sides of the one respectively equal to two sides of the other, and the angles opposite one pair of equal sides right angles, or equal obtuse angles, the triangles are congruent.*

For prop. XIV then applies ; the oblique angles cannot be supplemental.

6. *If two angles of one triangle equal two angles of another, the third angles are equal.*   (Why ?)

7. *Two triangles are congruent if two angles and any side of the one are respectively equal to the corresponding parts of the other.*   (Why ?)

8. *Each angle of an equilateral triangle is one-third of a straight angle.*   (Why ?)

89. **Definitions.**   A triangle, one of whose angles is a right angle, is called a **right-angled triangle.**

A triangle, one of whose angles is an obtuse angle, is called an **obtuse-angled triangle.**

A triangle, all of whose angles are acute, is called an **acute-angled triangle.**

The side opposite the right angle of a right-angled triangle is called the **hypotenuse.**

90. Summary of Propositions concerning Congruent Triangles. Two triangles are congruent if the following parts of the one are equal to the corresponding parts of the other :

1. *Two sides and the included angle.*   Prop. I

2. *Two angles and the included side.*   Prop. II

3. *Three sides.*   Prop. XII

4. *Two angles and the side opposite one,*   Prop. XIII
     or, more generally, *two angles and a side.*

Prop. XIX, cor. 7

5. *Two sides and the angle opposite one, provided that angle is not acute.*   Prop. XIV, and prop. XIX, cor. 5

If the angle is acute, then from two sides and the acute angle opposite one of them two different triangles may be possible. This is therefore known as the *ambiguous case.* If the side opposite the acute angle is not less than the given adjacent side, the case is not ambiguous. Why ? Draw the figures illustrating the ambiguous case.

These propositions can be summarized in one general proposition : *A triangle is determined when any three independent parts are given, except in the ambiguous case.*

It should be noted that the three angles are not three independent parts, since when any two of them are given the third is determined. (Prop. XIX.)

Exercises. 75. In a right-angled triangle, the mid-point of the hypotenuse is equidistant from the three vertices. (Suppose a line drawn from the vertex C of the right angle making with a an angle equal to $\angle B$.)

76. In a right-angled triangle, a perpendicular let fall from the vertex of the right angle, upon the hypotenuse, cuts off two triangles mutually equiangular to the original triangle.

77. If $a \perp x$ and $b \perp y$, and x intersects y, then $\angle ab = \angle xy$.

78. In the annexed figure, $\angle aa_1 = \angle bb_1$. Prove that
(1) $\angle aa_1 = \angle ab + \angle ba_1$; (2) $\angle bb_1 = \angle ba_1 + \angle a_1b_1$.

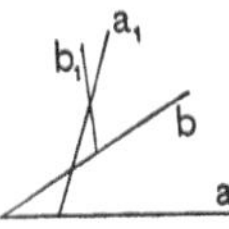

79. How many degrees in each angle of an isosceles right-angled triangle ? also of an isosceles triangle whose vertical angle is 72° ? 178° ? 60° ?

Proposition XX.

91. Theorem. *Of all lines drawn to a given line from a given external point, the perpendicular is the shortest; of others, those making equal angles with the perpendicular are equal; and of two others, that which makes the greater angle with the perpendicular is the greater.*

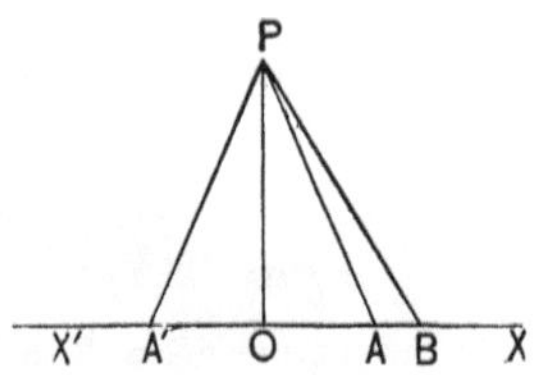

Given $PO \perp XX'$; PA, PA', PB, oblique to XX', with $\angle A'PO = \angle OPA < \angle OPB$.

To prove that

    (1) $PO < PA$,

    (2) $PA' = PA$,

    (3) $PB > PA$, or PA'.

Proof. 1. $\angle PAO < \angle AOP$.   Prop. XIX, cor. 1

    2. $\therefore PO < PA$, which proves (1).

               Prop. VII

    3. $\angle AOP = \angle POA'$,   Prel. prop. I

          $\angle A'PO = \angle OPA$,   Why ?

          and $PO \equiv PO$.

    4. $\therefore \triangle AOP \cong \triangle A'OP$,

          and $PA' = PA$, which proves (2). Why?

    5. $\angle BAP$ is obtuse, $\because$ it $> \angle AOP$,   Prop. V

      and $\angle PBO$ is acute, $\because \angle BOP$ is rt.   Why?

    6. $\therefore PB > PA$, or its equal PA',

      by step 4, which proves (3).   Prop. VII

Corollaries. 1. *From a given external point there can be two, and only two, equal obliques of given length to a given line.*

Prove it by a *reductio ad absurdum*.

2. *If from a point not on a perpendicular drawn to a line at its mid-point, lines are drawn to the ends of the line, these lines are unequal and the one cutting the perpendicular is the greater.*

Let Z be the point, not on OP in the figure. Suppose ZA' to cut OP at Y. Then $ZA' = ZY + YA > ZA$.

3. *The converse of cor. 2 is true.*

For if ZA' cuts OP, then $ZA' > ZA$, by cor. 2.

" ZA " " " " $<$ " " " "

" Z is on " " " $=$ " (Why ?)

∴ the Law of Converse (§ 73) evidently applies to this case.

4. *Of two obliques from a point to a line, that which meets the line at the greater distance from the foot of the perpendicular is the greater.*

For if $OB > OA$, then $\angle OPB > \angle OPA$. (Why ?) ∴ prop. XX applies.

5. *Two obliques from a point to a line, meeting that line at equal distances from the foot of the perpendicular, are equal, make equal angles with this line and also with the perpendicular.*

Give the proof in full.

6. *Two equal obliques from a point to a line cut off equal segments from the foot of the perpendicular.*

Draw the figure. It will then be seen that prop. XIX, cor. 5, applies. The ⊥ is evidently an axis of symmetry (§ 68).

Exercises. 80. A line perpendicular to the bisector of any angle of a triangle makes an angle with either arm of that angle equal to half the sum of the other two angles ; and, unless parallel to the base, it makes an angle with the line of the base equal to half the difference of those angles.

81. In an isosceles triangle, the perpendicular from the vertex, the median to the base, and the bisector of the vertical angle all coincide.

92. Definitions. A polygon is said to be **convex** when no side produced cuts the surface of the polygon.

A polygon is said to be **concave** when a side produced cuts the surface of the polygon.

A polygon is said to be **cross** when the perimeter crosses itself.

The word *polygon* is understood, in elementary geometry, to refer to a convex or concave polygon unless the contrary is stated.

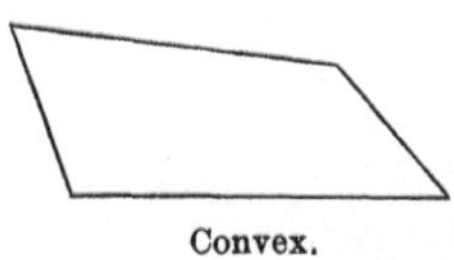
Convex.

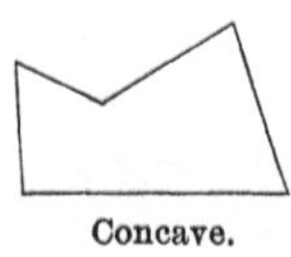
Concave.

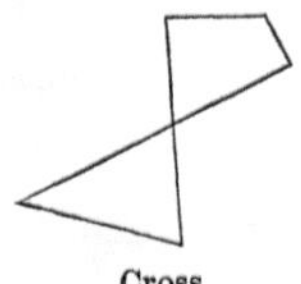
Cross.

If all of the sides of a polygon are indefinitely produced, the figure is called a **general polygon.**

If a polygon is both equiangular and equilateral, it is said to be **regular.**

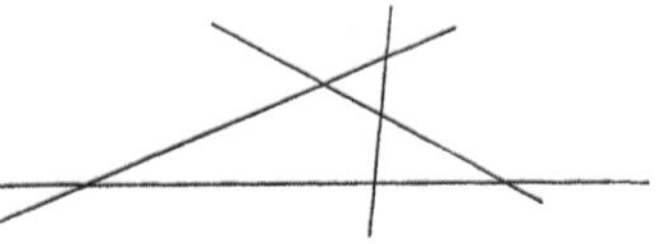
A general quadrilateral.

By the term *regular polygon*, a regular *convex* polygon is understood unless the contrary is stated.

A polygon is called a *triangle, quadrilateral, pentagon, hexagon, heptagon, octagon, nonagon, decagon,*

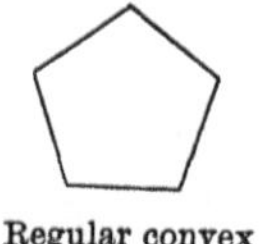
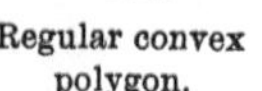
Regular convex polygon.

Regular cross polygon.

..... *dodecagon,* ..... *pentedecagon,* ..... *n-gon,* according as it has 3, 4, 5, 6, 7, 8, 9, 10, ... 12, ... 15, ... *n*-sides.

The student, even if unacquainted with Latin or Greek, should understand the derivation of these common terms. From the Latin are derived the words and prefix *tri-angle* (three-angle), *quadri-lateral* (four-side), *nona-* (nine); from the Greek are derived *poly-gon* (many-angle), *penta-* (five), *hexa-* (six), *hepta-* (seven), *octo-* (eight), *deca-* (ten), *dodeca-* (twelve), ... Much light will be thrown on the meaning of various geometric terms by consulting the Table of Etymologies in the Appendix.

Proposition XXI.

93. Theorem. *The sum of the interior angles of an n-gon is $(n-2)$ straight angles.*

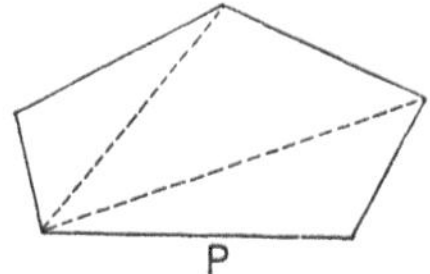

Given P, a polygon of n sides.

To prove that the sum of the interior angles is $(n-2)$ straight angles.

Proof. 1. P may be divided into $(n-2)$ △ by diagonals which do not cross; for,

(*a*) A 4-gon (quadrilateral) is a △ + a △,

∴ 2 △, or $(4-2)$ △.

(*b*) A 5-gon (pentagon) is a 4-gon + a △,

∴ 3 △, or $(5-2)$ △.

(*c*) A 6-gon (hexagon) is a 5-gon + a △,

∴ 4 △, or $(6-2)$ △.

(*d*) And every addition of 1 side adds 1 △.

(*e*) ∴ for an n-gon there are $(n-2)$ △.

2. The sum of the ∠s of each △ is a st. ∠. Prop. XIX

3. ∴ the sum of the interior ∠s of an n-gon is $(n-2)$ st. ∠s, because these equal the sum of the ∠s of the △.

Corollary. *If each of two angles of a quadrilateral is a right angle, the other two angles are supplemental.* (Why?)

Exercise. 82. How many diagonals in a common convex pentagon? hexagon?

94. Generalization of Figures. If a thermometer registers 70° above zero, it is ordinarily stated, in scientific works, that it registers $+70°$, while 10° below zero is indicated by $-10°$, the sign changing from $+$ to $-$ as the temperature decreases through zero. Similarly, west longitude is represented by the sign $+$, while longitude on the other side of 0° (*i.e.* east) is represented by the sign $-$, the longitude changing its sign in passing through zero. So in speaking of temperature it is said that $10° + (-10°) = 0$, meaning thereby that if the temperature rises 10° from 0, and then falls 10°, the result of the two movements is the original temperature, 0.

This custom holds in geometry. Thus, in this figure, if the segment between B and C is thought of as extending from B to C, it would be named BC; and, as is usually done in geometry with lines thought of as extending to the right, it would be considered a *positive* line.

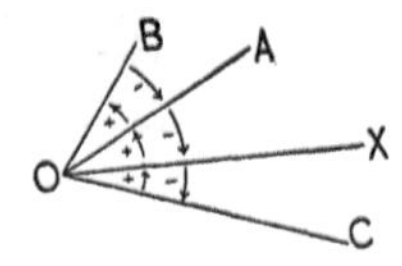

But if it is thought of as extending from C to B, it would be named CB, and considered a *negative* line. Hence it is said that $BC + CB = 0$, an expression borrowed from algebra, where it would appear in a form like $x + (-x) = 0$.

Similarly, with regard to angles: the turning of an arm in a sense opposite to that of a clock-hand, counter-clockwise, is considered *positive*, while the turning in the opposite sense is considered *negative*. Thus, $\angle XOA$ is considered positive, but the acute $\angle AOX$ is considered negative, and this is indicated by the statement, $-\angle XOA =$ acute $\angle AOX$. Hence, as in the case of lines, $\angle XOA + (-\angle XOA) = \angle XOA +$ acute $\angle AOX =$ zero. On this account we pay special attention to the manner of lettering angles, distinguishing between $\angle XOA$ and $\angle AOX$. It is only recently that negative angles have been considered in elementary geometry, and hence the older works paid no attention to the order of the naming of the arms.

95. These considerations enable us to *generalize* many figures, with interesting results.   Thus, prop. XXI is true for a cross polygon as well as for the simple cases usually considered.   If, in Fig. 1, P is moved through AB to the position

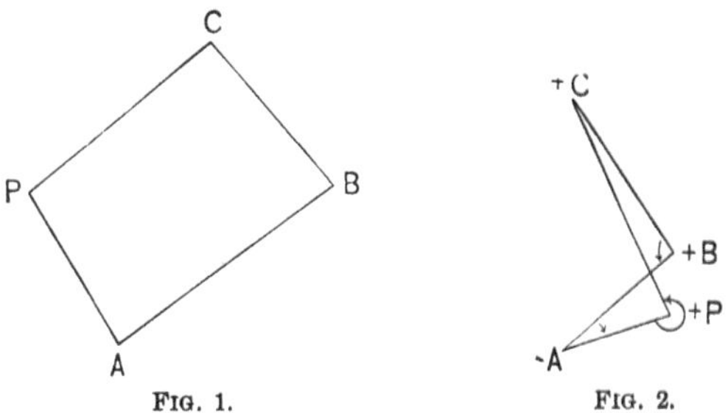

Fig. 1.   Fig. 2.

shown in Fig. 2, we shall still have $\angle A$ (which has passed through 0 and has become negative) $+ \angle B + \angle C + \angle P$ (which is now reflex) $= 2$ st. angles.

Exercises.   **83.**   Prove the last statement made above.

84.   How many points of intersection, at most, of the sides of a general quadrilateral ? pentagon ?

85.   How many diagonals, at most, has a general quadrilateral ?

86.   Prove prop. XXI by connecting each vertex with a point O within the figure, thus forming n ⧍, giving n st. ⧍, and then subtracting the two around O.

87.   In an isosceles triangle, the perpendiculars from the ends of the base to the opposite sides are equal.

88.   If the bisector of the vertical angle of a triangle also bisects the base, the triangle is isosceles.

89.   If the base AB of $\triangle ABC$ is produced to X, and if the bisectors of $\angle XBC$ and $\angle BAC$ meet at P, what fractional part is $\angle P$ of $\angle C$?

90.   Given two parallels and a transversal, what angle do the bisectors of the interior angles on the same side of the transversal make with each other ?

91.   If one angle of an isosceles triangle is given, and it is known whether it is the vertical angle or not, then the other two angles are determined.

Proposition XXII.

96. Theorem.   *The sum of the exterior angles of any polygon is a perigon.*

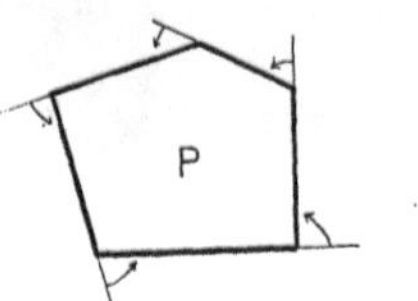 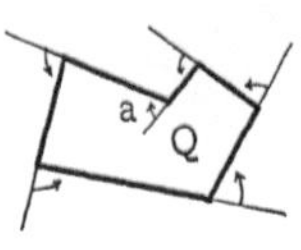

Given P and Q, two n-gons.

To prove   that the sum of the exterior $\angle$ = 360° in each n-gon.

Proof. 1. In P, each interior $\angle$ + its adjacent exterior $\angle$ = 180°.   § 14, def. st. $\angle$

2. $\therefore$ sum of int. and ext. $\angle$ = $n \cdot 180°$.   Ax. 6

3. But sum of int. $\angle$ = $(n - 2) \cdot 180°$.   Why ?

4. $\therefore$ sum of ext. $\angle$ = $2 \cdot 180°$ = 360°.   Ax. 3

The proof for Q is the same, if $\angle$ a is considered negative.

Exercises.   92. Each exterior angle of an equilateral triangle equals how many times each interior angle ?

93. Each exterior angle of a regular heptagon equals what fractional part of each interior angle ?

94. Each exterior angle of a regular n-gon equals what fractional part of each interior angle ?   See if the result found is true if $n = 3$, or 4.

95. Is it possible for the exterior angle of a regular polygon to be 70° ? 72° ? 75° ? 120° ?

96. Prove prop. **XXII** independently of prop. **XXI** by taking a point anywhere in the plane of the figure (inside or outside the polygon, or on the perimeter) and drawing parallels to the sides from that point, and showing that the sum of the exterior angles equals the perigon about that point.

97. Definitions. A quadrilateral whose opposite sides are parallel is called a **parallelogram.**

A quadrilateral that has one pair of opposite sides parallel is called a **trapezoid.**

Trapezium is a term often applied to a quadrilateral no two of whose sides are parallel.

By the definition of trapezoid here given it will be seen that the parallelogram may be considered a special form of the trapezoid.

The parallel sides of a trapezoid are called its *bases*, and are distinguished as *upper* and *lower*.

If the two opposite non-parallel sides of a trapezoid are equal, the trapezoid is said to be *isosceles*.

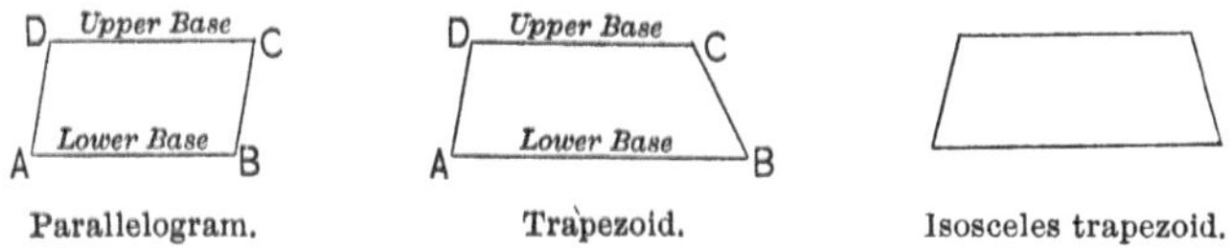

Parallelogram.   Trapezoid.   Isosceles trapezoid.

In the above figures, angles A, B, or B, C, or C, D, or D, A, are called *consecutive angles.* Angles A, C, or B, D, are called *opposite angles.*

Exercises. 97. If the student has proved ex. 96, let him prove prop. XXI from that.

98. Prove that the quadrilateral formed by the bisectors of the angles of any quadrilateral has its opposite angles supplemental.

99. Show that in ex. 98 the angles bisected may be either the four interior or the four exterior angles.

100. If from the ends of the base of an isosceles triangle perpendiculars are drawn to the opposite sides, a new isosceles triangle is formed, each of its base angles being half the vertical angle of the original triangle.

101. The hypotenuse is greater than either of the other sides of a right-angled triangle.

102. From the vertex of the right angle of a right-angled triangle, is it possible to draw, to the hypotenuse, a line longer than the hypotenuse? Proof.

103. A line from the vertex of an isosceles triangle to any point on the base produced is greater than either side. Is this also true for a scalene triangle?

Proposition XXIII.

98. Theorem. *Any two consecutive angles of a parallelo-gram are supplemental, and any two opposite angles are equal.*

Given $\square ABCD.$

To prove that　　(1) $\angle A + \angle B =$ st. $\angle$,

　　　　　　　　　(2)　　　　$\angle A = \angle C.$

Proof. 1.　　$\angle A + \angle B =$ st. $\angle$, which proves (1).

　　　　　　　　　　　　　　　　　　Prop. XVII, cor. 2

　　2.　　$\angle B + \angle C =$ st. $\angle.$　　　　　　　Why ?

　　3. $\therefore \angle A + \angle B = \angle B + \angle C.$　　　　Why ?

　　4.　　　$\therefore \angle A = \angle C$, which proves (2).　　Why ?

COROLLARY. *If one angle of a parallelogram is a right angle, all of its angles are right angles.* (Why ?)

99. Definitions. If one angle of a parallelogram is a right angle, the parallelogram is called a **rectangle**.

By the corollary, all angles of a rectangle are right angles.

A parallelogram that has two adjacent sides equal is called a **rhombus**.

It is shown in prop. **XXIV**, cor. 1, that all of its sides are equal.

A rectangle that has two adjacent sides equal is called a **square**.

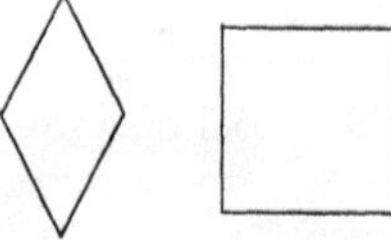
Rhombus.　　Square.

It is shown in prop. **XXIV**, cor. 1, that all of its sides are equal. A square is thus seen to be a special form of a rhombus.

Proposition XXIV.

100. Theorem. *In any parallelogram, (1) either diagonal divides it into two congruent triangles, (2) the opposite sides are equal.*

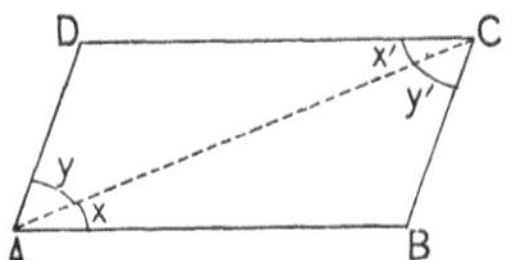

Given $\square\ ABCD.$

To prove   that   (1) $\triangle ABC \cong \triangle CDA,$

(2) $AB = DC.$

Proof. 1. In the figure, $\angle x = \angle x'$, $\angle y = \angle y'$, and $AC \equiv AC.$

Prop. (?)

2. $\therefore \triangle ABC \cong \triangle CDA$, which proves (1).   Prop. II

3. $\therefore AB = DC$, which proves (2).   § 57

Similarly for diagonal BD, and sides BC and $AD.$

Corollaries.   1. *If two adjacent sides of a parallelogram are equal, all of its sides are equal.*

For by step 3 the other sides are equal to these.
Hence, as stated in § 99, all of the sides of a rhombus are equal.

2. *The diagonals of a parallelogram bisect each other.*

For if diagonal BD cuts AC at O, then, by prop. II, $\triangle ABO \cong \triangle CDO$, whence $AO = OC$, and $BO = OD.$

In the annexed figure, if a and a' are perpendicular to P and P', two parallels (prop. XVII, cor. 1), they are parallel (prop. XVI, cor. 3).   Hence $a = a'$, by prop. XXIV.   This fact is usually expressed by saying,

3. *Two parallel lines are everywhere equidistant from each other.*

Proposition XXV.

101. Theorem. *If a convex quadrilateral has two opposite sides equal and parallel, it is a parallelogram.*

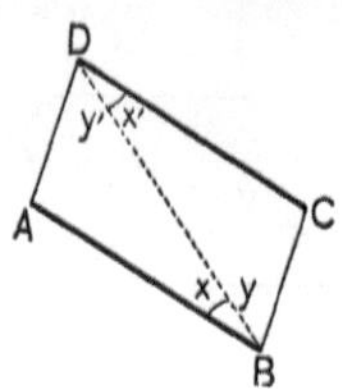

Given   a convex quadrilateral $ABCD$, with $AB = DC$, and $AB \parallel DC$.

To prove   that $ABCD$ is a parallelogram.

Proof.   1. In the figure $\angle x = \angle x'$,   Prop. (?)

          $BD \equiv BD$, and $AB = DC$.   Given

   2. $\therefore \triangle ABD \cong \triangle CDB$, and $\angle y = \angle y'$.   Prop. (?)

   3. $\therefore BC \parallel AD$.   Prop. XVI

   4. $\therefore ABCD$ is a $\square$ by definition.

Exercises.   **104.** It is shown in Physics that if two forces are pulling from the point B, and the first force is represented (see fig. to prop. XXV) by BA, and the second by BC, the resultant (resulting force) will be represented by the diagonal BD. Show that, if the two forces do not pull in the same line, the resultant is always less than the sum of the two forces.

105. If two equal lines bisect each other at right angles, what figure is formed by joining the ends?

106. If the diagonals of a rectangle are perpendicular to each other, prove that the rectangle is a square.

107. On the diagonal BD of $\square$ $ABCD$, P and Q are so taken that $BP = QD$. Show that $APCQ$ is a parallelogram. Suppose P is on DB produced, and Q on BD produced.

108. Prove that the diagonals of a rectangle are equal. Prove that the diagonals of a rhombus are perpendicular to each other and bisect the angles of the rhombus.

Proposition XXVI.

102. Theorem. *If two parallelograms have two adjacent sides and any angle of the one respectively equal to the corresponding parts of the other, they are congruent.*

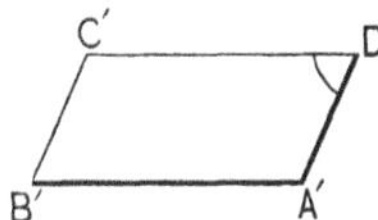 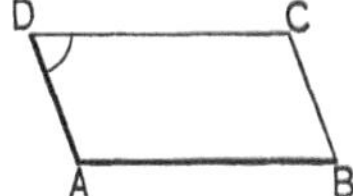

Given $\boxed{S}$ $ABCD$, $A'B'C'D'$, in which $AB = A'B'$, $AD = A'D'$, and $\angle D = \angle D'$.

To prove that $\Box\, ABCD \cong \Box\, A'B'C'D'$.

Proof. 1. $\angle A = \angle A'$, $\angle B = \angle B'$, $\angle C = \angle C'$, for they are equal or supplemental to D or D'.   Prop. XXIII

2. $CD = C'D'$, $BC = B'C'$, for they are equal to sides that are known to be equal.   Prop. XXIV

3. Apply $\Box\, ABCD$ to $\Box\, A'B'C'D'$ so that AB coincides with its equal $A'B'$, A falling on A'. Then AD can be placed on $A'D'$ because $\angle A = \angle A'$. Then D will fall on D', because $AD = A'D'$. Similarly, C will fall on C', and CB on $C'B'$.

Corollaries. 1. *Two rectangles are congruent if two adjacent sides of the one are equal to any two adjacent sides of the other.* (Why ?)

2. *Two squares are congruent if a side of the one equals a side of the other.* (Why ?)

Exercises. 109. Is a parallelogram determined when any two sides and either diagonal are given ? when two adjacent sides and either diagonal are given ?

110. The angles at either base of an isosceles trapezoid are equal.

Proposition XXVII.

103. Theorem. *If there are two pairs of lines, all of which are parallel, and if the segments cut off by each pair on any transversal are equal, then the segments cut off on any other transversal are equal also.*

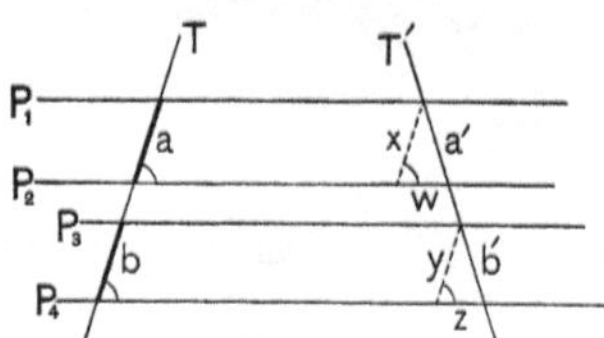

Given   four parallels, of which P_1, P_2 cut off a segment a, and P_3, P_4 cut off an equal segment b, on a transversal T, and cut off segments a', b', respectively, on transversal T'.

To prove that $a' = b'$.

Proof. 1. Suppose x and $y \parallel T$ as in the figure.

2. Then, in the figure, $\angle\, wx = \angle\, P_2 T = \angle\, P_4 T = \angle\, zy$.

    Prop. XVII, cor. 2

3. And $\angle\, a'w = \angle\, b'z$.   Why ?

4. And $x = a = b = y$.   Prop. XXIV

5. $\therefore \triangle\, wa'x \cong \triangle\, zb'y$, and $a' = b'$.   Prop. XIX, cor. 7

Corollaries. 1. *If a system of parallels cuts off equal segments on one transversal, it does on every transversal.*

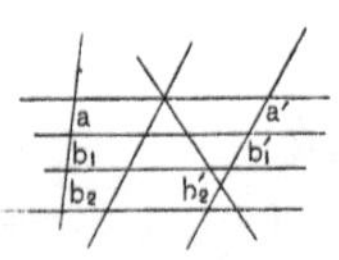

For if $a = b_1$ or b_2, $a' = b_1'$ or b_2', respectively, and similarly for the other transversals.

2. *The line through the mid-point of one side of a triangle, parallel to another side, bisects the third side.*

Draw a third parallel through the vertex. Then cor. 1 proves it.

3. *The line joining the mid-points of two sides of a triangle is parallel to the third side.*

For if not, suppose through the mid-point of one of those sides a line is drawn parallel to the base ; then this must bisect the other side, by cor. 2 ; ∴ it must coincide with the line joining the mid-points, or else a side would be bisected at two different points. (This is the converse of cor. 2.   Draw the figure.)

4. *The line joining the mid-points of two sides of a triangle equals half the third side.*   (Prove it.)

Exercises.   111. The line joining the mid-points of the non-parallel sides of a trapezoid is parallel to the bases.

112. In a right-angled triangle the mid-point of the hypotenuse is equidistant from the three vertices.   (This exercise has been given before, and will be repeated, since it is important and admits of divers proofs.   It is here easily proved by prop. XXVII, cor. 2 ; for if $a = b$, then $a' = b'$; but $p \parallel e$, ∴ $p \perp a'$, ∴ $x = b = a$.)

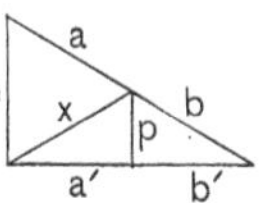

113. The lines joining the mid-points of the sides of a triangle divide it into four congruent triangles.

114. If one of the equal sides CB of an isosceles triangle ABC is produced through the base, and if a segment BD is laid off on the produced part, and an equal segment AE is laid off on the other equal side, then the line joining D and E is bisected by the base.   (Consider the cases in which $BD < CB$, $BD = CB$, $BD > CB$.)

115. If the mid-points of the adjacent sides of any quadrilateral are joined, the figure thus formed is a parallelogram.   (Consider this theorem for cases of concave, convex, and cross quadrilaterals, and for the special case of an interior angle of 180°.)

116. The lines joining the mid-points of the opposite sides of a quadrilateral bisect each other.   Consider for the special cases mentioned in ex. 115.

117. The line joining the mid-points of the diagonals of a quadrilateral, and the lines joining the mid-points of its opposite sides, pass through the same point.

118. P and Q are the mid-points of the sides AB and CD of the parallelogram $ABCD$.   Prove that PD and BQ trisect (divide into three equal segments) the diagonal AC.

EXERCISES.

119. What is the sum of the interior angles of a polygon of 20 sides ? of 30 sides ?

120. How many degrees in each angle of a regular polygon of 12 sides ? of 20 sides ?

121. How many sides has a polygon the sum of whose interior angles is 48 right angles ?

122. The vertical angle of a certain isosceles triangle is $11° 15' 20''$; how large are the base angles ?

123. The exterior angle of a certain triangle is 140°, and one of the interior non-adjacent angles is a right angle; how many degrees in each of the other two interior angles ?

124. Each exterior angle of a certain regular polygon is 10°; how many sides has the polygon ?

125. If P is any point on the side BC of $\triangle ABC$, then the greater of the sides AB, AC, is greater than AP.

126. If the diagonals of a quadrilateral bisect each other, prove that the quadrilateral is a parallelogram. Of what corollary is this the converse ? Prove that the diagonals of an isosceles trapezoid are equal.

127. Conversely, prove that if the diagonals of a trapezoid are equal, the trapezoid is isosceles.

128. Is a parallelogram determined when its two diagonals are given ? when its two diagonals and their angle are given ?

129. ABC is a triangle; AC is bisected at M; BM is bisected at N; AN meets BC at P; MQ is drawn parallel to AP to meet BC at Q. Prove that BC is trisected (see ex. 118) by P and Q.

130. A, C are points on the same side of XX'; B is the mid-point of AC; through A, B, C parallels are drawn cutting XX' in A', B', C'. Prove that $AA' + CC' = 2\,BB'$.

131. A straight line drawn perpendicular to the base AB of an isosceles triangle ABC cuts the side CA at D and BC produced at E; prove that CED is an isosceles triangle.

132. ABC is a triangle, and the exterior angles at B and C are bisected by the straight lines BD, CD respectively, meeting at D; prove that $\angle CDB + \frac{1}{2} \angle A =$ a right angle.

133. In the triangle ABC the side BC is bisected at E, and AB at G; AE is produced to F so that $EF = AE$, and CG is produced to H so that $GH = CG$. Prove that F, B, H are in one straight line.

3. PROBLEMS.

104. **Definitions.** A **curve** is a line no part of which is straight.

105. A **circle** is the finite portion of a plane bounded by a curve, which is called the **circumference,** and is such that all points on that line are equidistant from a point within the figure called the **center** of the circle.

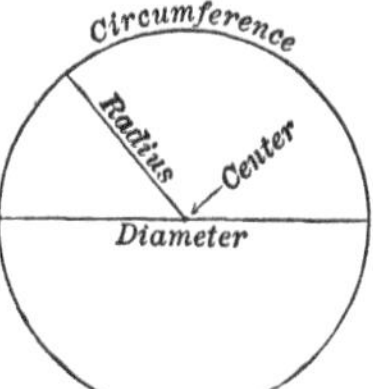

A circle is evidently described by a line-segment making a complete rotation in a plane, about a fixed point (the center).

106. A straight line terminated by the center and the circumference is called a **radius,** and a straight line through the center terminated both ways by the circumference is called a **diameter** of the circle.

107. A part of a circumference is called an **arc.**

NOTE. The above definitions are substantially those usually met in elementary geometries. The student will find, after leaving this subject, that the word *circle* is often used for *circumference.* Indeed, there is good authority for so using the word even in elementary geometry.

108. From the above definitions the following *corollaries* may be accepted without further proof :

1. *A diameter of a circle is equal to the sum of two radii of that circle.*

2. *Circles having the same radii are congruent.*

3. *A point is within a circle, on its circumference, or outside the circle, according as the distance from that point to the center is less than, equal to, or greater than, the radius.*

109. It now becomes necessary to assume certain postulates relating to the circle.

Postulates of the Circle.

1. *All radii of the same circle are equal, and hence all diameters of the same circle are equal.*

2. *If an unlimited straight line passes through a point within a circle, it must cut the circumference at least twice, and so for any closed figure.*

That it cannot cut the circumference more than twice is proved in III, prop. VI, cor.

3. *If one circumference intersects another once, it intersects it again.*

4. *A circle has but one center.*

5. *A circle may be constructed with any center, and with a radius equal to any given line-segment.*

This postulate requires the use of the compasses. As has been stated, *the only instruments allowed in elementary geometry are the compasses and the straight-edge*, a limitation due to Plato. In the more advanced geometry, where other curves than the circle are studied, other instruments are permitted.

110. ORDER TO BE OBSERVED in the solution of problems:

GIVEN. For example, the angle *A*.

REQUIRED. For example, to bisect that angle.

CONSTRUCTION. A statement of the process of solving, using only the straight-edge and compasses in drawing the figure described.

PROOF. A proof that the construction has fulfilled the requirements.

DISCUSSION. Any consideration of special cases, of the limitations of the problem, etc. If a problem has but a single solution, as that an angle may be bisected but once, the solution is said to be *unique*.

Proposition XXVIII.

111. Problem. *To bisect a given angle.*

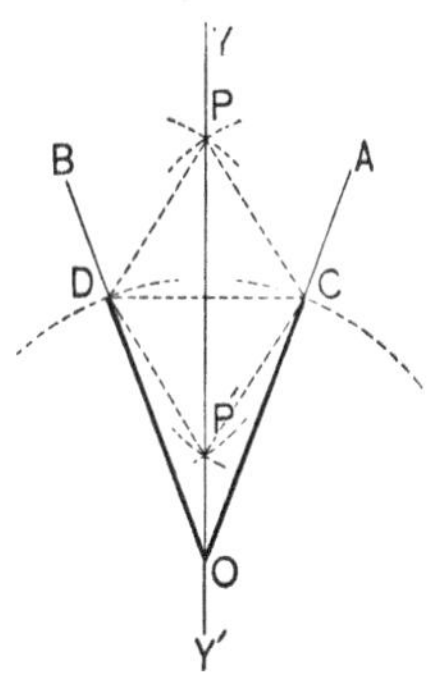

Given the $\angle AOB$.

Required to bisect it.

Construction. 1. With center O describe an arc cutting AO at C, and OB at D.   § 109, post. of ⊙

  2. Draw DC.   § 28, post. st. line

  3. Describe arcs with centers D, C, and radius DC.
        Post. (?)

  4. Join their intersection P, with O.   Post. (?)
     Then $\angle AOB$ is bisected, YY' being an axis of symmetry (§ 68).

Proof. 1. Draw DP, CP;
     then $OD = OC$, $DP = DC$, $DC = CP$.   § 109, 1

  2.   ∴ $DP = CP$.   **Ax.** (?)

  3. But $OP \equiv OP$.

  4.   ∴ $\triangle OCP \cong \triangle ODP$,
     and $\angle COP = \angle POD$.   Prop. XII

Corollary. *An angle may be divided into 2, 4, 8, 16, ...*
2^n, equal angles. (How ?)

112. Note on Assumed Constructions. It has been assumed, up to prop. XXVIII, that all constructions were made as required for the theorems. Thus an equilateral triangle has been frequently mentioned, although the method of constructing one has not yet been indicated; a regular heptagon has been mentioned in ex. 93, and reference might be made to certain results following from the trisection of an angle, although the solutions of the problems, to construct a regular heptagon, and to trisect any angle, are impossible by elementary geometry. But the possibility of solving such problems has nothing to do with the logical sequence of the theorems; one may know that each angle of a regular heptagon is $\frac{5}{7} \cdot 180°$, whether the regular heptagon admits of construction or not. Nevertheless, an important part of geometry concerns itself with the construction of certain figures — a part of utmost practical value and of much interest to the student of mathematics.

113. Suggestions on the Solution of Problems. The methods of logically undertaking the solution of problems will be discussed at the close of Book III. But at present one method, already suggested on p. 35, should be repeated: *In attempting the solution of a problem, assume that the solution has been accomplished; then analyze the figure and see what results follow; then reverse the process, making these results precede the solution.*

For example, in prop. XXVIII, *assume that* $\angle AOB$ *has been bisected by* YY'; *if that were done,* and if any point, P, on YY' were joined to points equidistant from O, on the arms, say C and D, then $\triangle OCP$ would be congruent to $\triangle ODP$; *now reverse the process* and attempt to make $\triangle OCP$ congruent to $\triangle ODP$; this can be done if OD can be made equal to OC, and PD to PC, because $OP \equiv OP$; but this can be done by § 109, 5.

This method of attacking a problem, without which the student will grope in the dark, is called **Geometrical Analysis.**

Exercises. 134. Give the solution of prop. XXVIII, using P' instead of P. Why is P better than P' for practical purposes? In what case would the construction fail for the point P'? In that case how many degrees in $\angle AOB$?

135. In prop. XXVIII, in what case would P' fall below O? Give the solution in that case, after connecting P' and O and producing $P'O$.

Proposition XXIX.

114. Problem. *To draw a perpendicular to a given line from a given internal point.*

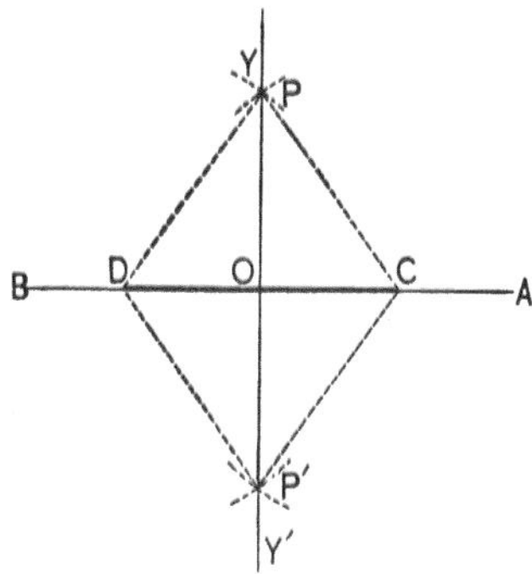

Solution. This is merely a special case of prop. XXVIII, the case in which $\angle AOB$ is a straight angle. (Why?) The construction and proof are identical with those of prop. XXVIII, and the student should give them to satisfy himself of this fact.

Exercises. 136. What kind of a quadrilateral is $CPDP'$? Prove it.

137. Prove that any point on BA is equidistant from P and P'. Also that any point on YY' is equidistant from D and C.

138. In step 3 of the construction of prop. XXVIII might the radius equal two times DC? If so, complete the solution. Is there any limit to the length of the radius in that step?

139. In the figure of prop. XXVIII, suppose $\angle PCO = 130°$. Find the number of degrees in the various other angles, not reflex, of the figure.

140. In the figure of prop. XXVIII, prove that the reflex angle BOA is bisected by YY', that is, by PO produced.

141. Also prove that YY' is the perpendicular bisector of DC.

142. Also prove that if O is connected with P and with P', OP' will fall on OP. (Prel. prop. VIII.)

Proposition XXX.

115. Problem. *To draw a perpendicular to a given line from a given external point.*

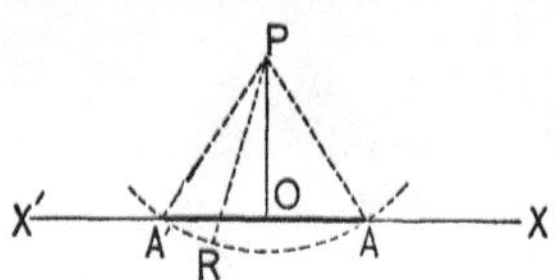

Given the line XX' and the external pt. P.

Required to draw a perpendicular from P to XX'.

Construction. 1. Draw PR cutting XX'.   § 28

    2. With center P and radius PR const. a $\odot$.

                     § 109, post. of $\odot$

    3. Join A and A', where the circumference cuts XX', with P.   § 28, post. of st. line

    4. Bisect $\angle A'PA$.   Prop. XXVIII

    The bisector, PO, is the required perpendicular.

Proof. 1. $PA = PA'$,   § 109, 1

                $\angle OPA = \angle A'PO$,   Const., 4

                and $PO \equiv PO$.

    2. $\therefore \triangle APO \cong \triangle A'PO$,   Prop. I

             and $\angle AOP = \angle POA'$.

    3. $\therefore \angle AOP$ is a rt. $\angle$, and $PO \perp XX'$.   §§ 19, 20

Nᴏᴛᴇ. The solution of this problem is attributed to Œnopides.

Exercises. 143. Find in a given line a point equidistant from two given points A and B, the mid-point of AB being also given.

144. Find a point equidistant from three given points A, B, C, the mid-points of AB and BC being also given.

Proposition XXXI.

116. Problem. *To bisect a given line.*

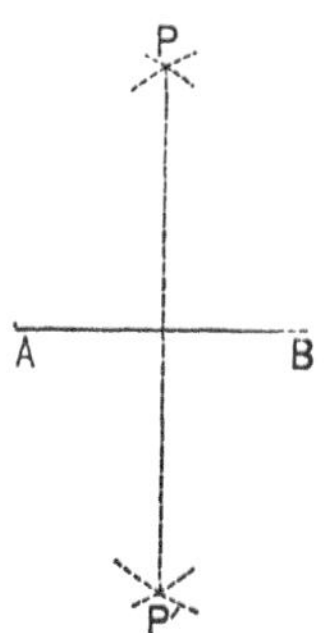

Given   the line *AB*.

Required to bisect it.

Construction. 1. With centers *A*, *B*, and equal radii describe
arcs intersecting at *P* and *P'*.   Post. (?)

2. Draw *PP'*.   Post. (?)

3. Then *PP'* bisects *AB*.

Proof.

(Let the student give it.   Draw *AP'*, *P'B*, *BP*, *PA*.)

Exercises. 145. Through a given point to draw a line making equal
angles with the arms of a given angle.   Discuss for various relative positions of the point.

146. To draw a perpendicular to a line from one of its extremities,
when the line cannot be produced.   (Ex. 112 suggests a plan.)

147. Through two given points on opposite sides of a given line draw
two lines which shall meet in the given line and include an angle which
is bisected by that line.

148. If two isosceles triangles have a common base, the straight line
through their vertices is a perpendicular bisector of the base.

Proposition XXXII.

117. Problem. *From a given point in a given line to draw a line making with the given line a given angle.*

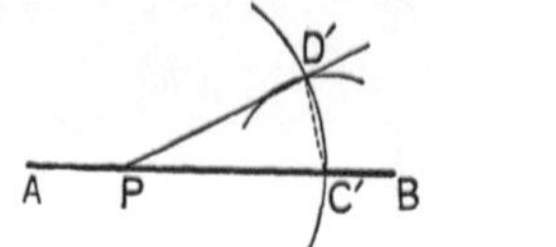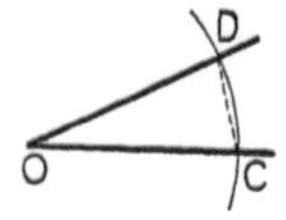

Given   the line AB, the point P in it, and the angle O.

Required from P to draw a line making with AB an angle equal to $\angle O$.

Construction. 1. On the arms of $\angle O$ lay off $OC = OD$ by describing an arc with center O and any radius OC.

§ 109, post. of ⊙

  2. Draw CD.   § 28, post. of st. line

  3. With center P and radius OC, describe a circumference cutting PB in C'.   Post. (?)

  4. With center C' and radius CD, describe an arc cutting the circumference in D'.   Post. (?)
Draw PD', and this is the required line.

Proof.   Draw $C'D'$; then,
$\triangle PC'D'$ and $\triangle OCD$ being mutually equilateral,

Why?

$$\triangle PC'D' \cong \triangle OCD,$$
$$\text{and } \angle C'PD' = \angle COD.\qquad \text{Prop. XII}$$

Exercises. 149. Prove that the circumferences must cut at D' as stated in step 4.

150. See if the solution of prop. XXXII is general enough to cover the cases where the $\angle O$ is straight, reflex, a perigon.

151. From a given point in a given line to draw a line making an angle supplemental to a given angle.

Proposition XXXIII.

118. Problem. *Through a given point to draw a line parallel to a given line.*

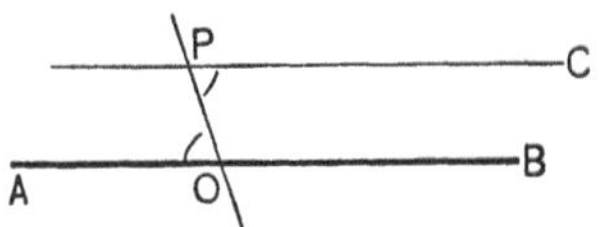

Given   the line AB and the point P.

Required   through P to draw a line parallel to AB.

Construction. 1. Join P with any point, O, on AB.

§ 28, post. of st. line

2. From P draw PC making $\angle OPC = \angle POA$.   (?)
Then PC is the required line.

Proof. $PC \parallel AB$.   Why?

Discussion. The solution fails if P is on the unlimited line AB.

Exercises. **152.** Through a given point to draw a line making a given angle with a given line.   Notice that the solution is not unique.

153. Through a given point to draw a transversal of two parallels, from which the parallels shall cut off a given segment.   Discussion should show when there are two solutions, when only one, when none.

154. To construct a polygon (say a hexagon) congruent to a given polygon.

155. Through two given points to draw two lines forming with a given unlimited line an equilateral triangle.

156. Three given lines meet in a point ; draw a transversal such that the two segments of it, intercepted between the given lines, may be equal. Is the solution unique ?

157. From P, the intersection of the bisectors of two angles of an equilateral triangle, draw parallels to two sides of the triangle, and show that these parallels trisect (see ex. 118) the third side.

Proposition XXXIV.

119. Problem. *To construct a triangle, given the three sides.*

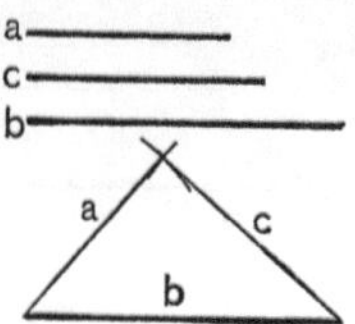

Given a, b, c, three sides of a triangle.

Required   to construct the triangle.

Construction. 1. With the ends of b as centers, and with radii a, c, describe circumferences.   § 109

     2. Connect either point of intersection of these circumferences with the ends of b.   § 28
     Then is the required $\triangle$ constructed.

Proof.   It was constructed on b, and the other sides equal a, c.
§ 109, 1

Discussion. If the two circumferences do not intersect, a solution is impossible, for then either $a > b + c$, $a = b + c$, $a = c - b$, or $a < c - b$, and in none of these cases is a triangle possible.

Prop. VIII and cor.

Corollary.   *To construct an equilateral triangle on a given line-segment.*

The first proposition of Euclid's "Elements of Geometry." Euclid proceeded upon the principle of logical sequence of propositions, with no attempt at grouping the theorems and the problems separately. He found this corollary (a problem) the best proposition with which to begin his system.

Exercise. 158. In a given triangle inscribe a rhombus, having one of its angles coincident with a given angle of the triangle, and the other three vertices on the three sides of the triangle.

Proposition XXXV.

120. Problem. *To construct a triangle, given two sides and the included angle.*

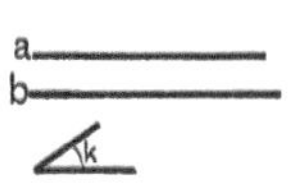 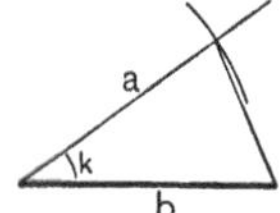

Given   the sides a, b, and the included angle k.

Required   to construct the triangle.

Construction. 1. From either end of b draw a line making with b the angle k.   Prop. XXXII

   2. On that line mark off a by describing an arc of radius a.   § 109

   3. Join the point thus determined with the other end of b.   § 28
   Then the triangle is constructed.

Proof.   By step 2 the line marked off equals a, and by step 1 $\angle b = \angle k$, and it is constructed on b.

Exercises. 159. To trisect a right angle. (Construct an equilateral triangle on one arm.)

160. On the side AC of $\triangle ABC$ to find the point P such that the parallel to AB, from P, meeting BC at D, shall have $PD = AP$.

161. To construct a triangle, having given two angles and the perpendicular from the vertex of the third angle to the opposite side.

162. Draw a line parallel to a given line, so that the segment intercepted between two other given lines may equal a given segment.

163. Given the three mid-points of the sides of a triangle, to construct the triangle.

164. Through a given point P in an angle AOB to draw a line, terminated by OA and OB, and bisected at P. (Through P draw a $\parallel$ to BO cutting OA in X; on XA lay off $XY = OX$; draw YP.)

Proposition XXXVI.

121. Problem. *To construct a triangle, given two sides and the angle opposite one of them.*

Given two sides of a triangle, *a*, *b*, and ∠ *k* opposite *a*.

Required to construct the triangle.

Construction. 1. At either end of *b* draw a line making with *b* an angle equal to ∠ *k*. Prop. (?)

2. With the other end of *b* as a center, and a radius *a*, describe a circumference. Post. (?)

3. Join the points where the circumference cuts the line of step 1, with the center. Post. (?)
Then the triangle is constructed.

Proof. For it has the given side *b*, and the given ∠ *k*, and the lines of step 3 equal *a*. § 109, 1

Discussion. If the circumference cuts the line twice, two solutions are possible, and the triangle is ambiguous (see prop. XIV). If it touches the line without cutting it, what about the solution? If it does not meet the line, no solution is possible. If ∠ *k* is right or obtuse, or if $a \not< b$, only one solution is possible (prop. XIX, cor. 5).

Draw a figure for each of these cases, and show from the drawings that the statements made in the discussion are true.

Exercise. **165.** XX', YY', are two given lines through *O*, and *P* is a given point; through *P* to draw a line to XX', which shall be bisected by YY'. Investigate for various positions of *P*, as where *P* is within the ∠ *XOY*, the ∠ *YOX'*, on *OY*, or on *OX*.

Proposition XXXVII.

122. Problem. *To construct a triangle, given two angles and the included side.*

Given   two angles, *A*, *B*, and the included side *AB*.

Required   to construct the triangle.

Construction.   1. From *A* draw *AX* making with *AB* an angle equal to ∠ *A*.   Prop. XXXII

    2. Similarly, from *B* draw *BY*, making an angle equal to ∠ *B*.   Prop. XXXII

    C being the intersection of *AX*, *BY*, then *ABC* is the required △.

Proof.   (Let the student give it.)

Discussion.   If *AX*, *BY*, do not intersect, what follows ?

Proposition XXXVIII.

123. Problem. *To construct a triangle, given two angles and a side opposite one of them.*

Solution.   Subtract the sum of the angles (found by prop. XXXII) from 180° and thus find the third angle (prop. XIX). The problem then reduces to prop. XXXVII.

Proposition XXXIX.

124. Problem. *To construct a square on a given line as a side.*

4. LOCI OF POINTS.

125. The place of all points satisfying a given condition is called the **locus of points** satisfying that condition.

Indeed, the word *locus* (Latin) means simply *place* (English, *locality, locate,* etc.) ; the plural is *loci.*

For example, if points are on this page and are one inch from the left edge, their locus is evidently a straight line parallel to the edge.

Furthermore, the locus of points at a given distance r from a fixed point O is the circumference described about O with a radius r. This statement, although very evident, is made a theorem (prop. XL) because of the frequent reference to it.

Of course in this discussion, as elsewhere in Books I–V, the points are all supposed to be confined to one plane.

In Plane Geometry the loci considered will be found to consist of one or more straight or curved lines.

It is a common mistake to assume that a locus, which one is trying to discover, consists of a single line. It may consist of two lines, as in prop. XLII.

126. In proving a theorem concerning the locus of points it is necessary and sufficient to prove two things :

1. *That any point on the supposed locus satisfies the condition;*

2. *That any point not on the supposed locus does not satisfy the condition.*

For if only the first were proved, there might be some other line in the locus ; and if only the second were proved, the supposed locus might not be the correct one.

Exercise. 166. State, without proof, what is (1) the locus of points $\frac{1}{4}$ in. from a given straight line ; (2) the locus of points equidistant from two parallel lines.

Proposition XL.

127. Theorem. *The locus of points at a given distance from a given point is the circumference described about that point as a center, with a radius equal to the given distance.*

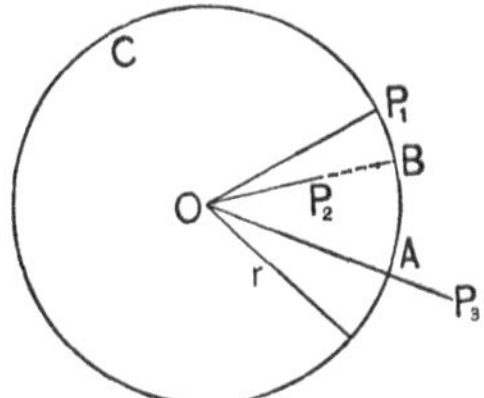

Given the point O, the line r, and the circumference C described about O with radius r.

To prove that C is the locus of points r distant from O.

Proof. 1. Let P_1, P_2, P_3 be points on C, within the circle, and without the circle, respectively.

Let OP_2 produced meet C in B, and OP_3 meet C in A.

2. Then $OP_1 = OB = OA = r$, § 109, 1

and $OP_2 < OB$, and $OP_3 > OA$. Ax. 8

3. ∴ any point on C is r distant from O, and any point not on C is not r distant from O.

Exercises. 167. Has it been proved in prop. XL that the required locus may not be merely the arc cut off by r and OP_1? If so, where?

168. What is the locus of points at a distance of $\frac{1}{4}$ in. from the above circumference, the distance being measured on a line through O?

169. Lighthouses on two islands are 10 miles apart; show that there are two points at sea which are exactly 12 miles from each.

170. How would you find, by the intersection of two loci, a point on this page 1 in. from O in the above figure, and 3 in. from the right edge of the paper?

PROPOSITION XLI.

128. Theorem. *The locus of points equidistant from two given points is the perpendicular bisector of the line joining them.*

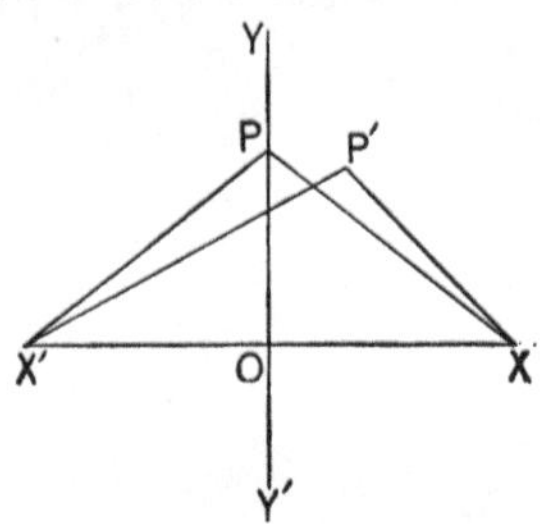

Given two points X and X', and $YY' \perp XX'$ at the midpoint O.

To prove that YY' is the locus of points equidistant from X and X'.

Proof. 1. Let P be any point on YY', and P' be any point not on YY'.

Draw PX, PX', $P'X$, $P'X'$.

2. Then $PX = PX'$,   Prop. XX, cor. 5

and $P'X' > P'X$.   Prop. XX, cor. 2

3. Hence any point on YY' is equidistant from X and X', but any point not on YY' is unequally distant from X and X'.

∴ YY' is the locus.   § 125, def.

Exercises. 171. Required to find a point which is 1 in. from X and $\frac{4}{2}$ in. from X' in the above figure. Is there more than one such point?

172. Required to find a point which is equidistant from X and X' in the above figure, and 1 in. from O. Is there more than one such point?

Proposition XLII.

129. Theorem. *The locus of points equidistant from two given lines consists of the bisectors of their included angles.*

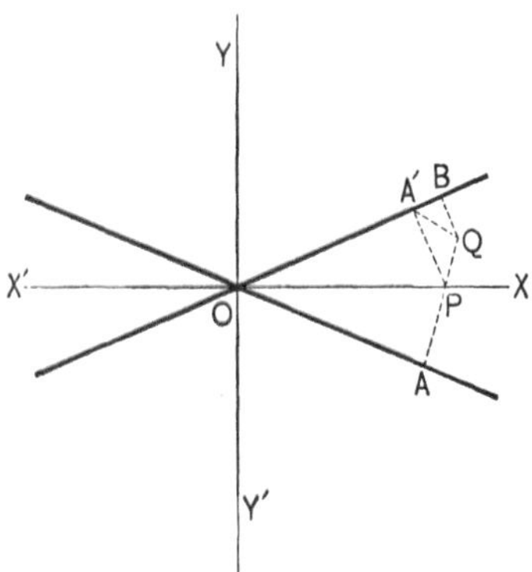

Given   *OA* and *OB*, two lines intersecting at *O*, and *XX'* and *YY'* the bisectors of the angles at *O*.

To prove   that *XX'* and *YY'* form the locus of points equidistant from *OA* and *OB*.

Proof. 1. Let *Q* be any point on neither *XX'* nor *YY'*; let $QB \perp OB$, $QA \perp OA$, QA cut *OX* in *P*, $PA' \perp OB$. Draw *QA'*.

Since *Q* may be moved, *P* may be considered as any point on *OX*.

2. Then $\triangle OAP \cong \triangle OA'P$,

and $AP = A'P$.   Prop. XIX, cor. 7

3. Also, $A'P + PQ > A'Q > BQ$.   Why ?

4. $\therefore AQ$, or $AP + PQ >$ BQ.   Why ?

5. $\therefore$ any point *P* on *XX'* (or on *YY'*) is equidistant from *OA* and *OB*, but any point *Q* on neither *XX'* nor *YY'* is unequally distant from *OA* and *OB*.

Corollaries. 1. *If the given lines are parallel, the locus is a parallel midway between them.* (Prove it.)

The student should imagine the effect of keeping points A, A' fixed, and moving O farther to the left. YY' moves with O, but XX' keeps its position as the lines approach the condition of being parallel.

2. *The locus of points at a given distance from a given line consists of a pair of parallels at that distance, one on each side of the fixed line.* (Prove it.)

130. Definitions.

Three or more *lines* which meet in a *point* are said to be **concurrent**.

Three or more *points* which lie in a *line* are said to be **collinear**.

Proposition XLIII.

131. Theorem. *The perpendicular bisectors of the three sides of a triangle are concurrent.*

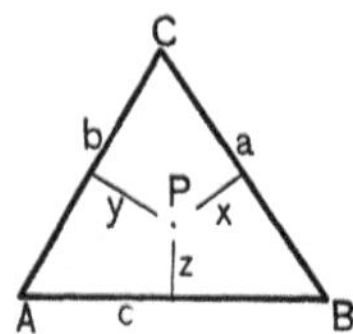

Given   a triangle of sides a, b, c, and x, y, z their respective perpendicular bisectors.

To prove   that x, y, z are concurrent.

Proof.   1. x and y must meet as at P.   Prop. XVII, cor. 4

        2.   Then P is equidistant from B and C, and C and A.

                                           Prop. XLI

        3. $\therefore$ P is on the perpendicular bisector of c;   Why?

           i.e. z passes through P.

Corollaries. 1. *The point equidistant from three non-collinear points is the intersection of the perpendicular bisectors of any two of the lines joining them.*

Step 2.

2. *There is one circle, and only one, whose circumference passes through three non-collinear points.*

Let A, B, C be the three points. Then by step 2 they are equidistant from P, the intersection of x and y.

And $\cdot\cdot$ x and y contain all points equidistant from A, B, and C, and can intersect but once, there is only one point P.

And $\cdot\cdot$ there is only one center and one radius, there is one and only one circle.

3. *Circumferences having three points in common are identical.*

Otherwise cor. 2 would be violated.

4. *If from a point more than two lines to a circumference are equal, that point is the center of the circle.*

For suppose a circumference through A, B, C, and suppose $PA = PB = PC$.

Now with center P and radius PA a circumference can be described through A, B, C, because it is given that

$$PA = PB = PC. \qquad\qquad \text{§ 108, cor. 3}$$

And this is identical with the given circumference.

Prop. XLIII, cor. 3

$\therefore$ its center must be identical with the given center, since a $\odot$ cannot have two centers.   § 109, 4

Exercises. 173. The proof of prop. XLIII is, of course, the same if the triangle is right-angled or obtuse-angled. The figures, however, show interesting positions for P; consider them.

174. Required to find a point at a given distance d from a fixed point O, and equidistant from two given intersecting lines. How many such points can be found in general ?

175. Required to find a point equidistant from two given intersecting lines, and equidistant from two given points. How many such points can be found in general ?

Proposition XLIV.

132. Theorem. *The bisectors of the interior and exterior angles of a triangle are concurrent four times by threes.*

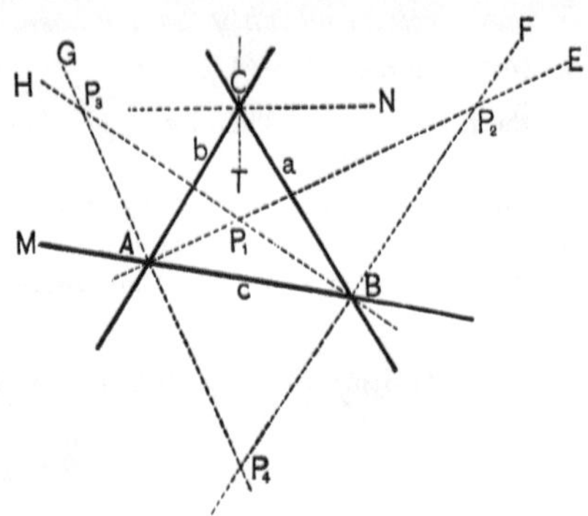

Given   the $\triangle ABC$, and the bisectors of the interior and exterior angles, lettered as in the figure.

To prove   that these bisectors are concurrent four times by threes; that is, 3 meet at P_1, 3 at P_2, etc.

Proof.   1. $\because \angle CAM > \angle CBM, \therefore \angle GAM > \angle HBM.$ Prop. V

2. $\therefore AG$ and BH meet as at P_3.   Prop. XVII, cor. 3

3. $\angle HBM + \angle BAE < \angle B + \angle A < 180°.$ Prop. XIX
$\therefore BH$ and AE meet as at P_1.   Prop. XVII, cor. 3

4. $BF \perp BH$, and $AG \perp AE$,   Prel. prop. IX
$\therefore BF$ and AG meet as at P_4.   Prop. XVII, cor. 4

5. Also, P_1 is equidistant from a and c, from c and b, and $\therefore$ from a and b,   Prop. XLII
$\therefore P_1$ lies on CT. Similarly for P_4.   Prop. XLII

6. Similarly, P_2 and P_3 lie on CN. $\therefore$ the four points P_1, P_2, P_3, P_4, are points of concurrence of the bisectors.

Proposition XLV.

133. Theorem. *The perpendiculars from the vertices of a triangle to the opposite sides are concurrent.*

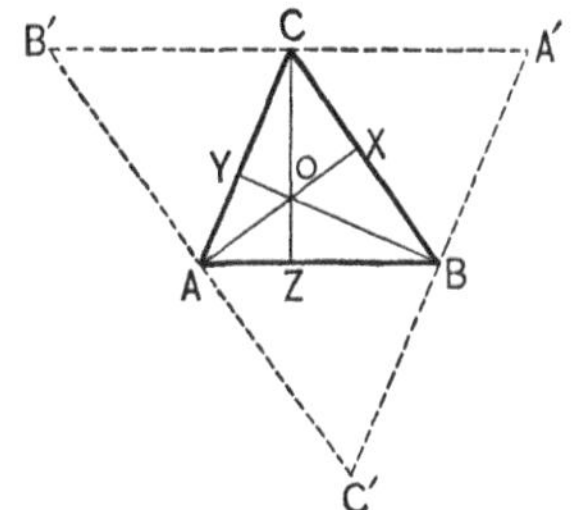

Given   the $\triangle ABC$.

To prove that the perpendiculars from A, B, C, to a, b, c, respectively, are concurrent.

Proof. 1. Through A, B, C, respectively, suppose $B'C' \parallel CB$, $A'C' \parallel CA$, $A'B' \parallel BA$.

    2. Then $ABCB'$ and $ABA'C$ are ▱.   Def. ▱

    3. $\therefore B'C = AB = CA'$, and C is the mid-point of $B'A'$.
                                     Prop. XXIV; ax. 1

    4. Similarly, A and B are mid-points of $B'C'$, $C'A'$.

    5. If AX, BY, $CZ \perp B'C'$, $C'A'$, $A'B'$, respectively, they are concurrent, as at O.   Prop. XLIII

    6. And they are also the perpendiculars from A, B, C to a, b, c.   Prop. XVII, cor. 1

Note. The theorem is due to Archimedes.

134. Definition. To **trisect** a magnitude is to cut it into three equal parts.

Exercise. 176. In prop. XLIV suppose C moves down to the side c. What becomes of P_1, P_2, P_3, P_4?

Proposition XLVI.

135. Theorem. *The medians of a triangle are concurrent in a trisection point of each.*

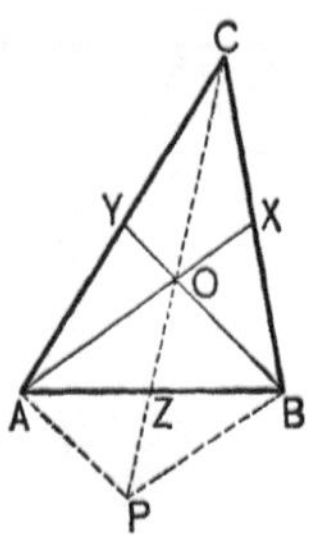

Given   the $\triangle ABC$ and the medians BY, AX, intersecting at O.

To prove that (1) the median from C must pass through O,
(2) $OX = \frac{1}{3} AX$, $OY = \frac{1}{3} BY$, etc.

Proof. 1. Suppose CO drawn, and produced indefinitely, cutting AB at Z.

2. Suppose $AP \parallel OB$; CO must cut AP, as at P.   § 85

3. Draw PB.   Then $\because CY = YA$, $\therefore CO = OP$.
Prop. XXVII, cor. 2

4. And $\because CO = OP$, and $CX = XB$, $\therefore OX \parallel PB$.
Prop. XXVII, cor. 3

5. $\therefore APBO$ is a $\square$, $AZ = ZB$, and $OZ = ZP$.
Prop. XXIV, cor. 2

6. $\therefore CZ$ is a median, and it passes through O.

7. And $\because OZ = \frac{1}{2} OP$, $\therefore OZ = \frac{1}{2} CO$, or $\frac{1}{3} CZ$.   Similarly for OY and OX.

Exercise.   177. The sum of the three medians of a triangle is greater than three-fourths of its perimeter.

136. Definitions. The point of concurrence of the perpendicular bisectors of the sides of a triangle is called the **circumcenter** of the triangle. (Prop. XLIII.)

The reason will appear later when it is shown that this point is the *center* of the *circum*-scribed circle. (See Table of Etymologies.)

137. The point of concurrence of the bisectors of the *interior* angles of a triangle is called the **in-center** of the triangle; the points of concurrence of the bisectors of two *exterior* angles and one interior are called the **ex-centers** of the triangle. (Prop. XLIV.)

It will presently be proved that the *in*-center is the center of a circle, *in*-side the triangle, just touching the sides; and that the *ex*-centers are centers of circles, *out*-side the triangle, just touching the three lines of which the sides of the triangle are segments. Hence the names *in-center* and *ex-center*.

138. The point of concurrence of the three perpendiculars from the vertices to the opposite sides is called the **orthocenter** of the triangle. (Prop. XLV.)

139. The point of concurrence of the three medians of a triangle is called the **centroid** of that triangle. (Prop. XLVI.)

It is shown in Physics that this point is also the center of mass, or center of gravity of the plane surface of the triangle. It is, therefore, sometimes called by those names.

Exercises. 178. If a triangle is acute-angled, prove that both the circumcenter and the orthocenter lie within the triangle.

179. In prop. XLVI, if X, Y, Z be joined, prove that the $\triangle XYZ$ will be equiangular with the $\triangle ABC$.

180. Is there any kind of a triangle in which the in-center, circumcenter, orthocenter, and centroid coincide? If so, what is it? Prove it.

181. In the figure of prop. XLVI, connect X, Y, Z, and prove that O is also the centroid of $\triangle XYZ$.

182. In ex. 179, prove that if the mid-points of the sides of $\triangle XYZ$ are joined, O is also the centroid of that triangle; and so on.

1. THEOREMS.

140. Definitions. Two polygons are said to be **adjacent** if they have a segment of their perimeters in common.

141. Suppressing the common segment of the perimeters of two adjacent polygons, a polygon results which is called the **sum** of the two polygons. Similarly for the *sum* of several polygons, and for the **difference** of two overlapping polygons.

142. Surfaces which may be divided into the same number of parts respectively congruent, or which are the differences between congruent surfaces, are said to be **equal.**

This property is often designated by the expressions *equivalent, equal in area, of equal content,* etc.; but the use of the word *congruent,* for *identically equal,* renders. the word *equal* sufficient.

The definition is more broadly treated in Book V.

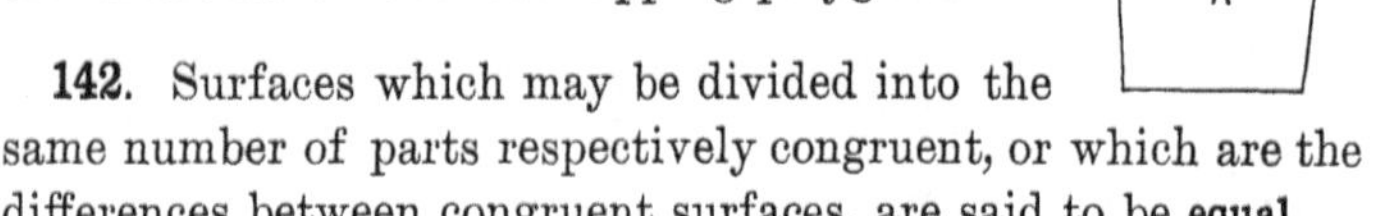

143. The **altitude of a trapezoid** is the perpendicular distance between the base lines.

Hence a trapezoid can have but one altitude, a, unless it becomes a parallelogram.

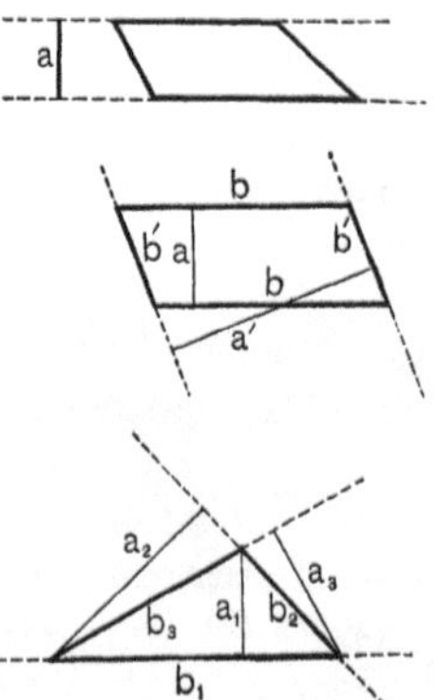

144. The **altitude of a triangle** with reference to a given side as the base, is the distance from the opposite vertex to the base line.

Hence a triangle can have three distinct altitudes, viz. a_1, a_2, a_3, in the figure.

Proposition I.

145. Theorem. *Parallelograms on the same base or on equal bases and between the same parallels are equal.*

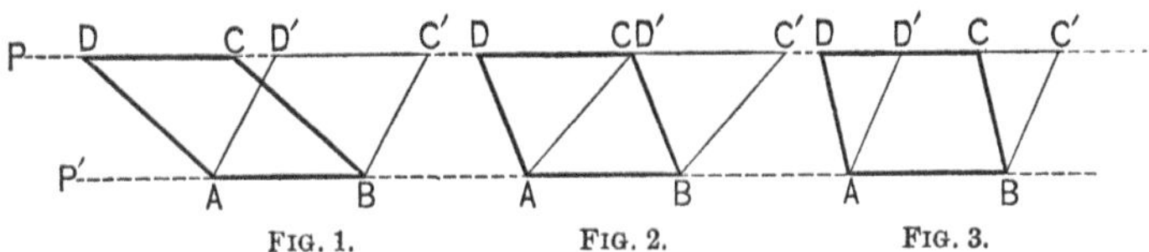

Given $\boxed{s}$ $ABCD$, $ABC'D'$, on the same base AB, and between the same parallels P, P'.

To prove   that $\square\, ABCD = \square\, ABC'D'$.

Proof. 1. $AD = BC$, $AD' = BC'$, $DC = AB = D'C'$.   Why?

2. In Fig. 1, adding CD', $DD' = CC'$.   Ax. 2

3. $\therefore \triangle BC'C \cong \triangle AD'D$.   Why?

4. But $ABC'D \equiv ABC'D$.

$\therefore \square\, ABCD = \square\, ABC'D'$.   Ax. 3

Similarly for Figs. 2 and 3.   In Fig. 2, CD' has become zero; in Fig. 3, it has become negative.

The meaning of "between the same parallels" is apparent.

Corollaries.   1. *A parallelogram equals a rectangle of the same base and the same altitude.*   (Why?)

2. *Parallelograms having equal bases and equal altitudes are equal.*   (Why?)

3. *Of two parallelograms having equal altitudes, that is the greater which has the greater base; and of two having equal bases, that is the greater which has the greater altitude.*   (Why?)

4. *Equal parallelograms on the same base or on equal bases have equal altitudes.*

Law of Converse, § 73, after cors. 2 and 3.   Give it in full.

Proposition II.

146. **Theorem.** *Triangles on the same base or on equal bases and between the same parallels are equal.*

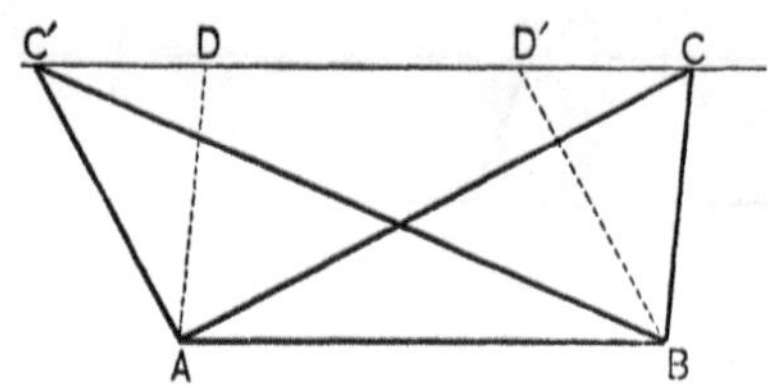

Given   ⚠ *ABC, ABC'* on the base *AB*, and between the same parallels *AB, C'C*.

To prove that   △ *ABC* = △ *ABC'*.

Proof. 1. In the figure, suppose *AD* ∥ *BC, BD'* ∥ *AC'*.

Then *ABCD, ABD'C'* are equal ▱.   Why ?

2. And, since ⚠ *ABC, ABC'* are their halves,

I, prop. XXIV

∴ △ *ABC* = △ *ABC'*.   Ax. 7

Corollaries.   1. *A triangle equals half of a parallelogram, or half of a rectangle, of the same base and the same altitude as the triangle.*

By step 2, and prop. I, cor. 1.

2. *Triangles having equal bases and equal altitudes are equal.*

3. *Of two triangles having equal altitudes, that is the greater which has the greater base ; and of two having equal bases, that is the greater which has the greater altitude.*   (Why ?)

4. *Equal triangles on the same base or on equal bases have equal altitudes.*   (Why ?)

Note.   In props. I and II if the figures are on *equal* bases they can evidently be placed on the same base. Hence the proofs given are sufficient.

Proposition III.

147. Theorem. *A trapezoid is equal to half of the rectangle whose base is the sum of the two parallel sides, and whose altitude is the altitude of the trapezoid.*

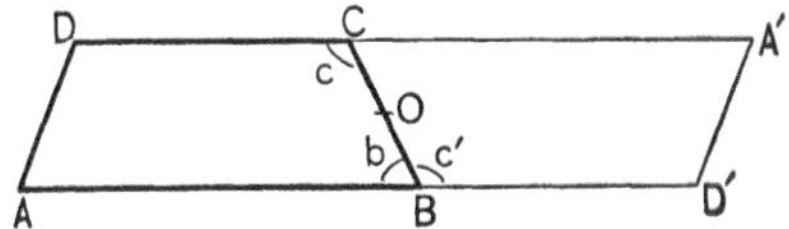

Given   the trapezoid $ABCD$.

To prove that $ABCD$ equals half of a rectangle with the same altitude, and with base equal to $AB + DC$.

Proof. 1. About O, the mid-point of BC, revolve $ABCD$ through $180°$ to the position $A'CBD'$, leaving its original trace.

  2. Then, $\because \angle c' \equiv \angle c$, and $\angle b + \angle c = $ st. $\angle$,

    $\therefore \angle b + \angle c' = \angle b + \angle c = $ st. $\angle$,

    and $\therefore ABD'$ is a st. line.   § 14, def. st. $\angle$

    Similarly, DCA' is a st. line.

  3. Also, $\because \angle D' \equiv \angle D$, and $\angle A + \angle D = $ st. $\angle$,

    $\therefore \angle A + \angle D' = \angle A + \angle D = $ st. $\angle$,

    $\therefore D'A' \parallel AD$,   I, prop. XVI, cor. 2

    and $\therefore AD'A'D$ is a $\square$.   § 97, def. $\square$

  4. The base of the $\square = AB + DC$,

    and the $\square = 2 \cdot ABCD$.   Why?

  5. $\therefore ABCD = \tfrac{1}{2}\,\square = \tfrac{1}{2}$ required $\square$.   Prop. I, cor. 1

Exercises. 183. P is any point within $\square$ $ABCD$. Prove that $\triangle PAB + \triangle PCD = \tfrac{1}{2} \square ABCD$. Suppose P is outside of $\square$ $ABCD$.

184. A quadrilateral equals a triangle of which two sides equal the diagonals of the quadrilateral, and the included angle of those sides equals the included angle of the diagonals.

PROPOSITION IV.

148. Theorem. *If through a point on a diagonal of a parallelogram parallels to the sides are drawn, the parallelograms on opposite sides of that diagonal are equal.*

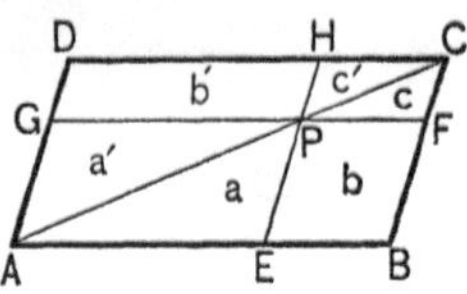

Given $\square\ ABCD$, and through P, a point on AC, the lines $GF \parallel AB$, $HE \parallel DA$, and the parts lettered as in the figure.

To prove that $b = b'$.

Proof. 1. $a + b + c = a' + b' + c',$

             $a = a', \quad c = c'.$ I, prop. XXIV

     2. $\therefore b = b'.$ Ax. 3

149. Definitions. Since all rectangles which have two adjacent sides equal to two given lines a, b, are congruent (I, prop. XXVI, cor. 1), any such rectangle is spoken of as **the rectangle of *a* and *b*.**

This is indicated by the symbol ab, or, if the adjacent sides are AB and CD, by the symbol $AB \cdot CD$. These symbols are read "The rectangle of a and b," "The rectangle of AB and CD," or, briefly, "The rectangle ab," "The rectangle AB (pause) CD." Since there is no multiplication of lines by lines, by any definition thus far known to the student, the readings "a *times* b," "AB *times* CD" are not recommended.

In like manner, any square whose side is equal to a given line is spoken of as **the square on (or of) that line.**

The square on a line AB is indicated by the symbol AB^2; on a line a by the symbol a^2; read "The square on (or of) AB," or, briefly, "AB-square"; and similarly for a.

Squares, rectangles, and polygons in general are often designated by the letters of two vertices not consecutive.

150. A point in a line-segment is said to **divide it internally;** a point in a produced part of a line-segment is said to **divide it externally.**

In the figure, AB is divided internally at P, and externally at P'. AP, PB are called *segments* of AB; and AP', $P'B$ are also called segments of AB.

The propriety of calling AP, PB, and AP', $P'B$, segments of AB is apparent, since $AP + PB = AB$, and also $AP' + P'B$ (which is negative) $= AB$.

Exercises. 185. If the sides BC, CA, AB, of $\triangle ABC$, are produced to X, Y, Z, respectively, so that $CX = BC$, $AY = CA$, $BZ = AB$, prove that $\triangle XYZ = 7 \cdot \triangle ABC$.

186. The medians of a triangle divide it into six equal triangles. (In what kind of a triangle are the six triangles congruent?)

187. Prove prop. III by bisecting BC at O, drawing DO to meet AB produced at D', and proving that $\triangle BD'O \cong \triangle CDO$, that $\triangle AD'D =$ trapezoid, etc.

188. Discuss prop. IV when P moves to C; through C on AC produced.

189. If two equal triangles are on opposite sides of a common base the line of that base bisects the line joining their vertices.

190. A triangle X is equal to a fixed triangle T and has a common base with T; what is the locus of the vertex of X? (Is the locus a single line or a pair of lines?)

191. P is any point on the diagonal BD of $\square ABCD$. Prove that $\triangle PAB = \triangle PBC$.

192. In ex. 191, suppose P moves to D; moves through D on BD produced.

193. The sides AB, CA of a triangle are bisected in C', B', respectively; CC' cuts BB' at P. Prove that $\triangle PBC =$ quadrilateral $AC'PB'$.

194. If P is a point on the side AB, and Q a point on the opposite side CD of $\square ABCD$, prove that $\triangle PCD = \triangle QAB$.

195. If the mid-points of the sides of any convex quadrilateral are joined, in order, then (1) a parallelogram is formed, (2) which equals half the quadrilateral.

Proposition V.

151. Theorem. *The rectangle of two given lines equals the sum of the rectangles contained by one of them and the several segments into which the other is divided.*

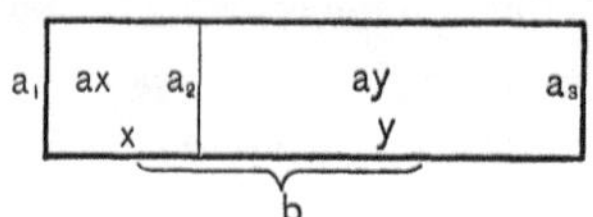

Given the rectangle of a_1 and b, and b divided into the segments x and y.

To prove that $a_1b = a_1x + a_1y.$

Proof. 1. Let a_2 be drawn $\parallel a_1$ from the division point of b.

2. Then, in the figure, $a_1 = a_2 = a_3$.   I, prop. XXIV

3. $\therefore a_1b = a_1x + a_1y.$ Ax. 8

CorOLLARIES. **1.** *If a line is divided internally into two segments, the rectangle of the whole line and one segment equals the square on that segment together with the rectangle of the two segments.*

Make $a_1 = x$ in the proof above, and consider step 3.

2. *If a line is divided internally into two segments, the square on the whole line equals the sum of the rectangles of the whole line and each of the segments.*

Make $a_1 = x + y$ in the proof above.

Note. This theorem is the geometric form of the **Distributive Law of Multiplication** of Algebra, which asserts that $a(x + y) = ax + ay.$

Exercise. 196. The rectangle of one line and the sum of two others equals the sum of the rectangles of the first and each of the other two. Consider the case where one of the second two lines is zero ; negative.

152. **Positive and Negative Polygons.**   In general, a line AB is thought of as positive; but if, in the discussion of a proposition, A is thought of as approaching B, then, when A reaches B, AB becomes zero; and if A is thought of as passing through B, then AB is considered as having passed through zero and become negative; that is, $BA = -AB$.

A similar agreement exists as to triangles. In general, $\triangle ABC$ is thought of as positive; but if, in the discussion of a proposition, C moves down to rest on AB, then $\triangle ABC$ becomes zero; and as C passes through AB, $\triangle ABC$ passes through zero

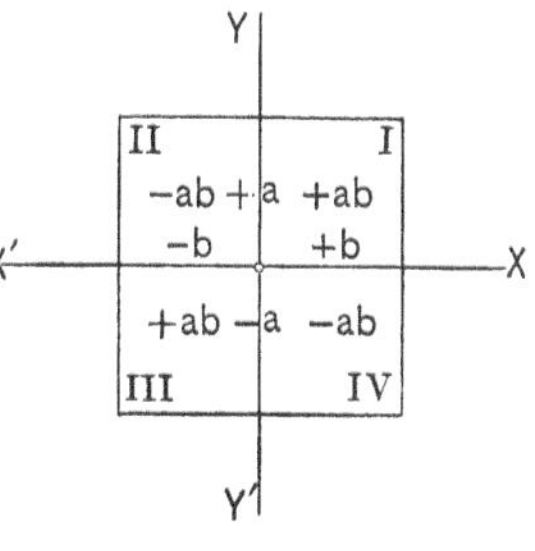

and is considered as having changed its sign and become negative; that is, $\triangle ACB = -\triangle ABC$.

In Book I, to accustom students to this convention (that $\triangle ACB = -\triangle ABC$), triangles were always named by taking the letters in the counter-clockwise (or positive) order, except in a few cases where a departure from this rule seemed advisable.

A similar agreement exists *as to rectangles*, which illustrates the law of signs in algebra.   In the figure, I has for its altitude and base $+a$ and $+b$, and the rectangle is spoken of as $+ab$. But if b shrinks to zero, $+ab$ also shrinks to zero, and as b passes through zero and becomes negative, so ab is considered to pass through zero and to become negative; that is, $II = -ab$.   If, now, a shrinks to zero, and passes through zero, changing its sign, so does $-ab$; that is, $III = +ab$.   And finally, as $-b$ again passes through zero, so does ab, and therefore $IV = -ab$.

Exercise. **197.** If P is any point in the plane of $\triangle ABC$, then $\triangle PAB + \triangle PBC + \triangle PCA = \triangle ABC$.   (Monge.)

In the case of polygons in general, the law of signs will be readily understood from the annexed figures. In Figs. 1, 2, 3 both the upper and lower parts of the polygon are considered as positive; in Fig. 4, P

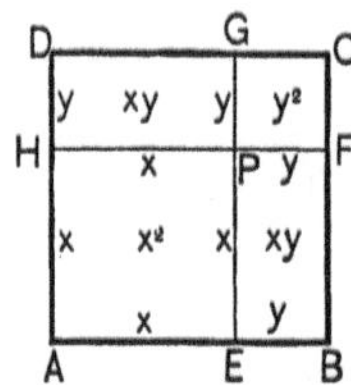

has reached BC and the upper part of the polygon has become zero; in Fig. 5, P has passed through BC and the upper part of the figure has passed through zero and become negative.

This treatment of negative surfaces dates from Meister (1769).

PROPOSITION VI.

153. Theorem. *The square on the sum of two lines equals the sum of the squares on those lines plus twice their rectangle.*

Given $ABCD$, the square on $x + y$.

To prove that $(x + y)^2 = x^2 + y^2 + 2\,xy.$

Proof. 1. In the figure, let $AE = AH = x$, $EB = HD = y$, $EG \parallel AD$, $HF \parallel AB$, and HF cut EG at P.

     2. Then the x's in the figure are all equal; also the y's.
                              Def. $\square$; I, prop. XXIV

     3. $\therefore AP = x^2$, $PC = y^2$,   § 99, def. $\square$
       and $EF = xy$, $HG = xy$.   § 99, def. $\square$

     4. $\therefore (x + y)^2 = x^2 + y^2 + 2\,xy.$ Ax. 8

COROLLARIES. 1. *The square on a line equals four times the square on half that line.*

Make $x = y$ in step 4.

Then $$(2\,x)^2 = x^2 + x^2 + 2\,x^2,$$

or $$(2\,x)^2 = 4\,x^2.$$

That is, if $2\,x$ is the line, the square upon it equals four times the square on x.

2. *The square on the difference of two lines equals the sum of the squares on those lines minus twice their rectangle.*

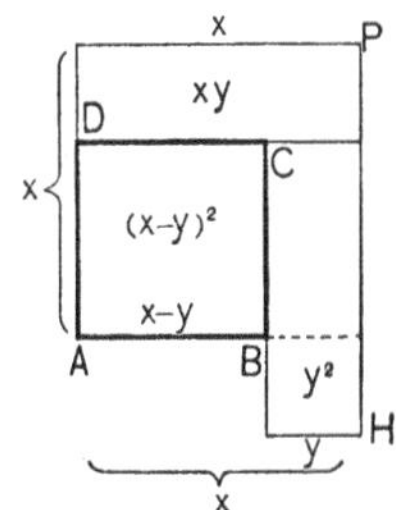

For, in the above figure, $AP = x^2$, $BH = y^2$, $PD = xy$, $CH = xy$, and $AC = (x - y)^2$.

But $$AC = AP + BH - PD - CH,$$

or $$(x - y)^2 = x^2 + y^2 - 2\,xy.$$

The truth of the corollary is, however, evident from prop. VI, if the agreement as to signs is considered ; for if y becomes 0, then $2\,xy = 0$, and $y^2 = 0$; and as y passes through 0 and becomes negative, $2\,xy$ also becomes negative, but y^2 remains positive because it is the rectangle of $-y$ and $-y$.

Exercises. 198. If $ABC \ldots\ldots MNA$ is the perimeter of any polygon, and P is any point in the plane, then $\triangle\,PAB + \triangle\,PBC + \ldots\ldots + \triangle\,PMN + \triangle\,PNA$ is constant.

199. If A, B, C, D are four collinear points, in order, then $AB \cdot CD + AD \cdot BC = AC \cdot BD$. (Euler.) Investigate when B moves to and through C.

Proposition VII.

154. Theorem. *The difference of the squares on two lines equals the rectangle of the sum and difference of those lines.*

Given $ABCD$, a square on x, and $AEFG$, a square on y.

To prove that $x^2 - y^2 = (x + y)(x - y)$.

Proof. 1. Suppose the squares placed as in the figure, and GF produced to BC. Then the y's in the figure are all equal, as also the sides of x^2.   Def. □

  2. ∴ the $(x - y)$'s are equal.   Ax. 3

  3. But $x^2 = y^2 + x(x - y) + y(x - y)$,   Ax. 8

    or $x^2 = y^2 + (x + y)(x - y)$.   Prop. V

  4. ∴ $x^2 - y^2 = (x + y)(x - y)$.   Ax. 3

Cᴏʀᴏʟʟᴀʀʏ. *If a point divides a line internally or externally into two segments, the rectangle of the segments equals the difference of the square on half the line and the square on the segment between the mid-point of the line and the point of division.*

1. If AB is the line, P (either P_1 or P_2) the point of division, and M the mid-point, it is to be proved that

$$AP \cdot PB = AM^2 - MP^2.$$

2. Let $AB = y$, $AP = x$, then $PB = y - x$, $AM = \frac{1}{2}y$, and $MP = x - \frac{1}{2}y$.

3. But $x(y - x) = \left(\dfrac{y}{2}\right)^2 - \left(x - \dfrac{y}{2}\right)^2$, by prop. VII.

155. Reciprocity between Algebra and Geometry. From props. V, VI, VII, it is evident that a reciprocity exists between algebra and geometry which is likely to be of great advantage to each. This reciprocity will be more clearly seen by resorting to parallel columns.

<table>
<tr><td>

Geometric Theorems.

If x, y, ... are *line-segments*, and xy, xz, ... represent the *rectangles* of x and y, x and z, ..., and $x(y+z)$ represents the *rectangle* of x and $y+z$, and x^2 represents the *square* on x, then

1. $x(y+z) = xy + xz.$
    Prop. V

2. $(x+y)^2 = x^2 + y^2 + 2xy.$
    Prop. VI

3. $\qquad x^2 = 4\left(\dfrac{x}{2}\right)^2.$
    Prop. VI, cor. 1

4. $(x-y)^2 = x^2 + y^2 - 2xy.$
    Prop. VI, cor. 2

5. $x^2 - y^2 = (x+y)(x-y).$
    Prop. VII

</td><td>

Algebraic Theorems.

If a, b, ... are *numbers*, and ab, ac, ... represent the *products* of a and b, a and c, ..., and $a(b+c)$ represents the *product* of a and $b+c$, and a^2 represents the *second power* of a, then

1. $a(b+c) = ab + ac.$

2. $(a+b)^2 = a^2 + b^2 + 2ab.$

3. $\qquad a^2 = 4\left(\dfrac{a}{2}\right)^2.$

4. $(a-b)^2 = a^2 + b^2 - 2ab.$

5. $a^2 - b^2 = (a+b)(a-b).$

</td></tr>
</table>

This correspondence of one symbol, one operation, one result, etc., of algebra, to one symbol, one operation, one result, etc., of geometry, or, as it is called, this *"one-to-one correspondence,"* suggests many theorems of geometry that might otherwise remain unnoticed. This correspondence is the basis of the treatment of Proportion, Book IV.

Exercise. 200. Prove geometrically that $(x+y)^2 - (x-y)^2 = 4xy.$

PROPOSITION VIII.

156. Theorem. *In a right-angled triangle the square on the hypotenuse equals the sum of the squares on the other two sides.*

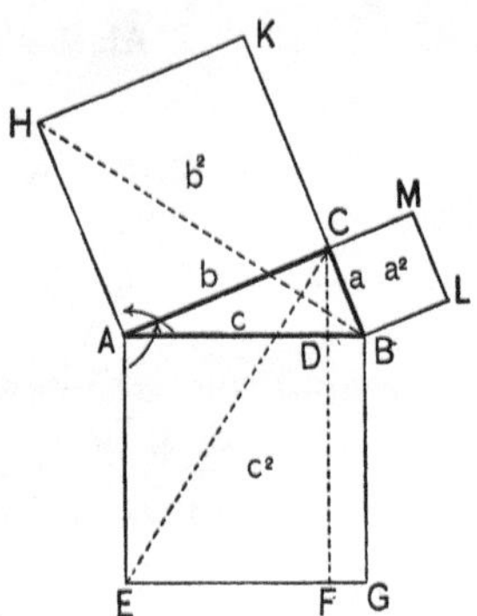

Given the right-angled $\triangle ABC$, $\angle C$ being right.

To prove that $a^2 + b^2 = c^2$.

Proof. 1. Let $BLMC = a^2$, $ACKH = b^2$, $AEGB = c^2$; suppose $CF \parallel AE$, and HB and CE drawn.

   2. Then $\because \triangle KCA$, ACB, are right, their sum $=$ st. $\angle$, and $\therefore BCK$ is a st. line.   § 14, def. st. $\angle$

   3. And $\because \angle CAH = \angle EAB$,   Prel. prop. I

     $\therefore \angle BAH = \angle EAC$, by adding $\angle BAC$.   Ax. 2

   4. And $\because AC = HA$, and $AE = BA$,   § 99, def. $\square$

     $\therefore \triangle ABH \cong \triangle AEC$.   Why ?

   5. But $\square AF = $ twice $\triangle AEC$,

      and $b^2 = $ twice $\triangle ABH$.   Why ?

     $\therefore \square AF$, or $AB \cdot AD$, $= b^2$.   Ax. 6

   6. Similarly, $\square BF$, or $BA \cdot BD$, $= a^2$.

   7. $\therefore a^2 + b^2 = \square AF \quad + \square BF = c^2$.   Axs. 2, 8

157. NOTE. The first proof of this theorem is said to have been given by Pythagoras about 540 B.C., although the theorem itself was known long before that time. From this fact it is generally known as the **Pythagorean Proposition.** It is one of the most important in geometry.

There have been many proofs devised for the Pythagorean proposition. In the subsequent exercises occasional proofs will be suggested, that the student may see the great variety of ways in which the theorem may be attacked. That the proposition would naturally be suggested to a people using tile floors is seen from Fig. 1, although the proof following

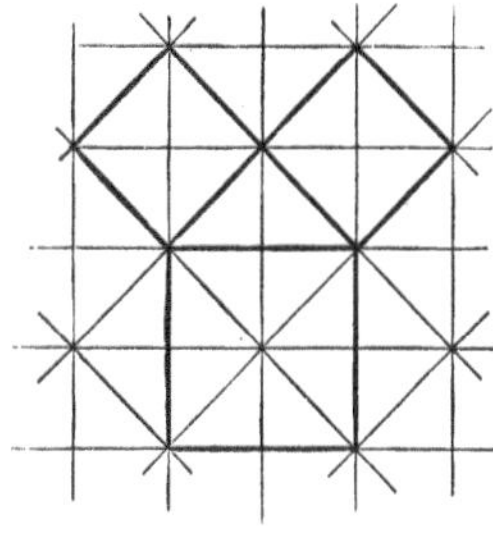

FIG. 1.

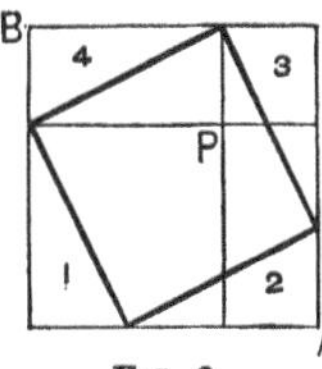

FIG. 2.

from such a figure is special, being limited to the case of the isosceles right-angled triangle.

In Fig. 2 is given a suggestion of the conjectured proof of Pythagoras: If $\triangle$ 1, 2, 3, 4 are taken from the figure, the square on the hypotenuse remains; and if the two $\square$ *AP*, *PB*, are taken away, the sum of the squares on the two sides remains; but since the two rectangles equal the four triangles, these remainders are equal.

Exercises. 201. What is the use of steps 2 and 3 in the proof of prop. VIII ?

202. In the figure of prop. VIII, prove that $AK \parallel BM$.

203. Also that *H, C, L* are collinear.

204. Twice the sum of the squares on the medians of a right-angled triangle equals thrice the square on the hypotenuse.

Fig. 3 is that of Bhaskara, the Hindu: The inside square is evidently $(a - b)^2$, and each of the four triangles is $\frac{1}{2}ab$; $\therefore c^2 - 4 \cdot \frac{1}{2}ab = (a - b)^2 = a^2 + b^2 - 2ab$; $\therefore c^2 = a^2 + b^2$.

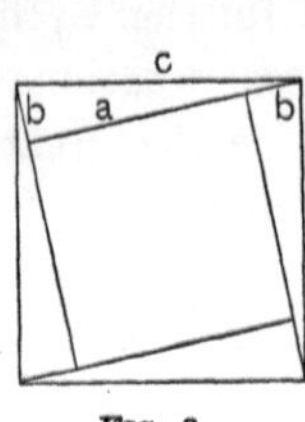

FIG. 3.

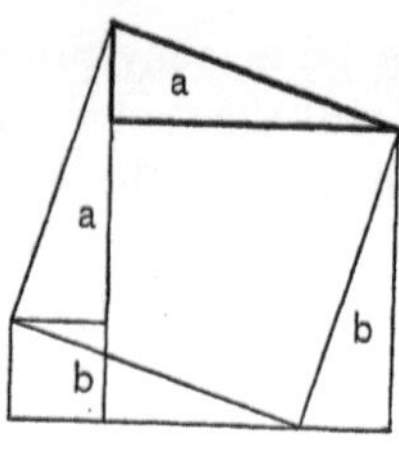

FIG. 4.

Fig. 4 is one of the most simple: If from the whole figure there are taken $\triangle b$, there remains the square on the hypotenuse; or if the equal $\triangle a$ are taken, there remains the sum of the squares on the two sides.

158. Definition. The **projection of a point on a line** is the foot of the perpendicular from the point to the line.

Thus A' and B', Figs. 1, 2, are the projections of A and B on $X'X$.

The **projection of a line-segment** on another line in the same plane is the segment cut off by the projections of its end-points, *e.g.* in Figs. 1 and 2, $A'B'$ is the projection of AB.

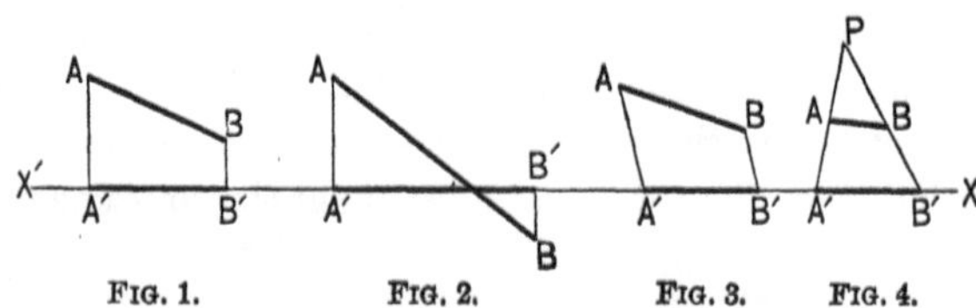

FIG. 1.　　　FIG. 2.　　　FIG. 3.　　　FIG. 4.

Strictly these are *orthogonal* (or right-angled) projections; but since orthogonal projections are the only kind ordinarily considered in elementary geometry, they are called, simply, *projections.* In advanced geometry, the projections of Figs. 3, 4 are among the others used. Fig. 3 represents an oblique projection, and Fig. 4 represents a projection from a point. Fig. 4 is the most general, approaching the others as P recedes to a greater distance.

Proposition IX.

159. Theorem. *In an obtuse-angled triangle the square on the side opposite the obtuse angle equals the sum of the squares on the other two sides, together with twice the rectangle of either side and the projection of the other on the line of that side.*

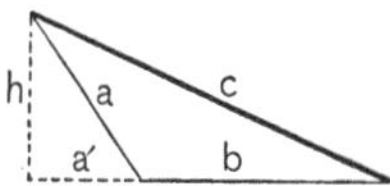

Given $\triangle\,abc$, obtuse-angled opposite c, and a' the projection of a on the line of b.

To prove that $c^2 = a^2 + b^2 + 2\,ba'$.

Proof. 1. In the figure, $h \perp a'$,   § 158, def. projection

$$\therefore h^2 + (a' + b)^2 = c^2,\qquad \text{Why ?}$$

or $h^2 + a'^2 + b^2 + 2\,a'b = c^2$.   Prop. VI

2. $\therefore a^2 + b^2 + 2\,a'b = c^2$.   Prop. VIII

Exercises. 205. In the figure of prop. VIII, prove that, if HK and LM are produced to meet at P, then $AE = $ and $\parallel PC$, and $BG = $ and $\parallel PC$.

206. If the diagonals of a quadrilateral intersect at right angles, prove that the sum of the squares on one pair of opposite sides equals the sum of the squares on the other pair.

207. In the annexed figure, equilateral triangles are constructed on the sides of a right-angled triangle; M is the mid-point of CA. Prove (1) $\triangle\,ABK \cong \triangle\,AEC$, (2) $MK \parallel BC$, (3) $\triangle\,BCM = \triangle\,BCK$, (4) $\triangle\,BRM = \triangle\,RCK$, (5) $\triangle\,ABK = \triangle\,ACK + \triangle\,ABM = \triangle\,ACK + \tfrac{1}{2}\triangle\,ABC$, (6) $\therefore$ from (1) and (5) $\triangle\,AEC = \triangle\,ACK + \tfrac{1}{2}\triangle\,ABC$, (7) similarly, $\triangle\,CEB = \triangle\,BLC + \tfrac{1}{2}\triangle\,ABC$, (8) $\therefore$ figure $AEBC = $ figure $ABLCK$, (9) $\therefore \triangle\,AEB = \triangle\,BLC + \triangle\,ACK$. State in full form the theorem proved in (9).

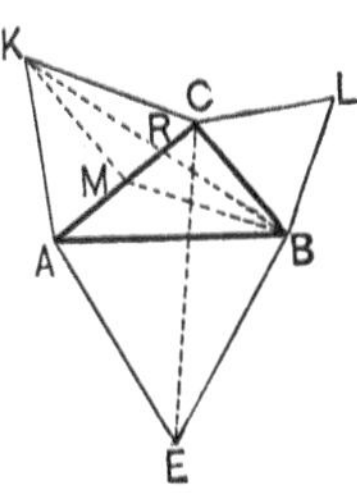

Corollaries. 1. *In any triangle the square on the side opposite an acute angle equals the sum of the squares on the other two sides, less twice the rectangle of either side and the projection of the other side on it.*

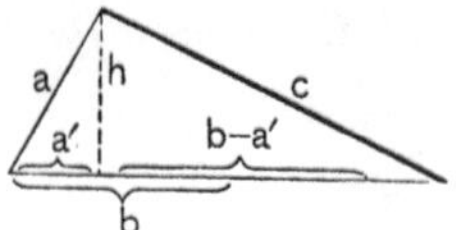

For, in the above figure,

$$h^2 + (b - a')^2 = c^2.$$
$$\therefore h^2 + b^2 + a'^2 - 2\,a'b = c^2.$$
$$\therefore a^2 + b^2 - 2\,a'b = c^2.$$

The truth of the corollary is, however, evident from prop. IX; for if $\angle ba$ becomes 90°, $a' = 0$ and prop. IX becomes prop. VIII; and if $\angle ba$ becomes acute, a' passes through 0 and becomes negative, and $\square\,a'b$ becomes negative; $\therefore$ step 2 becomes $a^2 + b^2 - 2\,a'b = c^2$.

2. *Converse of props. VIII, IX, and prop. IX, cor. 1.   The angle opposite a given side of a triangle is right, obtuse, or acute, according as the square on that side is equal to, greater or less than the sum of the squares on the other two sides.*

Law of Converse (§ 73).   Write out the proof in full.

Exercises. 208. In the figure of prop. VIII, the medians of $\triangle\,ABC$ are perpendicular to and equal to half of KM, HE, LG, respectively. (Complete the $\square\,BCAV$, and prove $CV =$ and $\perp KM$, etc.)

209. XOY is any angle, and from B, on OY, BA is drawn $\perp OX$; from B is drawn $BZ \parallel OX$; now *if* P can be found on BA, so that OP produced to cut BZ in Q, makes $PQ = 2\,OB$, then $\angle XOQ = \frac{1}{3}\angle XOY$. (That is, $\angle XOY$ is trisected.   It has been *proved* that this famous problem of the Greeks, to trisect any angle, cannot be solved by elementary geometry, that is, by using the compasses and straight-edge only.   There are various solutions if other instruments are allowed.)

210. Prove algebraically that if n is an even number, then n, $\frac{1}{4}n^2 - 1$, $\frac{1}{4}n^2 + 1$ are numerically the sides of a right-angled triangle (Plato), and that they are integers.

Proposition X.

160. Theorem. *The sum of the squares on any two sides of a triangle equals twice the sum of the squares on one-half the third side and on the median to that side.*

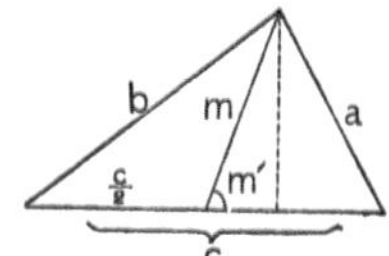

Given   the $\triangle abc$, and m the median to c.

To prove that $a^2 + b^2 = 2\left[\left(\dfrac{c}{2}\right)^2 + m^2\right].$

Proof. 1. Let m' be the projection of m on c, and suppose $\angle cm$ acute.

2. Then $a^2 = \left(\dfrac{c}{2}\right)^2 + m^2 - 2\left(\dfrac{c}{2}\right)m',$ Prop. IX, cor. 1

   and $b^2 = \left(\dfrac{c}{2}\right)^2 + m^2 + 2\left(\dfrac{c}{2}\right)m'.$ Why?

3. $\therefore a^2 + b^2 = 2\left[\left(\dfrac{c}{2}\right)^2 + m^2\right].$ Ax. 2

   If $\angle cm$ is obtuse, then $\angle mc$ is acute, and the proof merely interchanges a, b without affecting step 3. If $\angle cm$ is right, then $m' = 0$ in step 2, but 3 is not affected.

Exercises. 211. In prop. X, prove that $4m^2 = 2(a^2 + b^2) - c^2$. Hence show that in a right-angled triangle (in which $a^2 + b^2 = c^2$) the median to the hypotenuse equals half the hypotenuse.

212. From ex. 211, what is the locus of the vertex of the right angle of a right-angled triangle with a given hypotenuse?

213. The sides of a triangle are 10, 12, 15 inches. Is the triangle right-angled? obtuse-angled?

Proposition XI.

161. Theorem. *The sum of the squares on the sides of a quadrilateral equals the sum of the squares on the diagonals plus four times the square on the line joining the mid-points of the diagonals.*

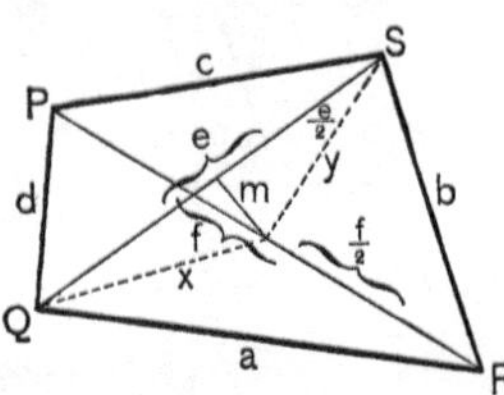

Given a quadrilateral *abcd*, convex, concave, or cross, with diagonals e, f, and with m joining the mid-points of e, f.

To prove that $a^2 + b^2 + c^2 + d^2 = e^2 + f^2 + 4\,m^2$.

Proof. 1. In the figure, $a^2 + d^2 = 2\,x^2 + 2\left(\dfrac{f}{2}\right)^2$,   Why?

and $b^2 + c^2 = 2\,y^2 + 2\left(\dfrac{f}{2}\right)^2.$ Why?

2. $\therefore a^2 + b^2 + c^2 + d^2 = 2\,(x^2 + y^2) + 4\left(\dfrac{f}{2}\right)^2$ Ax. 2

$$= 2\left[2\left(\frac{e}{2}\right)^2 + 2\,m^2\right] + 4\left(\frac{f}{2}\right)^2.$$

Prop. X

$$= e^2 + f^2 + 4\,m^2. \quad \text{Prop. VI, cor. 1}$$

Corollary. *The sum of the squares on the diagonals of a parallelogram equals the sum of the squares on the sides.*

For then $m = 0$; I, prop. XXIV, cor. 2.

Noᴛᴇ. The theorem is due to Euler. The corollary was, however, known to the Greeks.

2. PROBLEMS.

Proposition XII.

162. Problem. *To construct a triangle equal to a given polygon.*

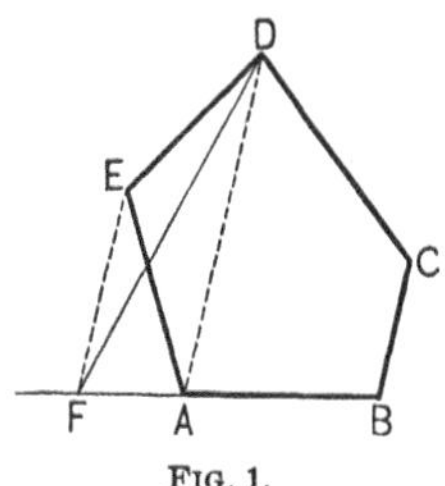

Given   polygon $ABCDE$.

Required to construct a $\triangle$ equal to $ABCDE$.

Construction. 1. Produce BA, join D and A, draw $EF \parallel DA$, meeting BA produced at F; draw DF.

§ 28, post. of st. line; I, prop. XXXIII

   2. Then polygon $FBCD$ has one less side than $ABCDE$, and will be proved equal to it. Continue the process until a $\triangle$ is reached (Fig. 2).

Proof. 1. $\because EF \parallel DA$,

   $\therefore \triangle ADF = \triangle ADE$, having same base AD.

Prop. II

   2. Adding polygon $ABCD$, $FBCD = ABCDE$.   Ax. 2

   3. Similarly thereafter. In Fig. 2, $\triangle FGD$ is the triangle required.

Exercise. 214. To construct a rhombus equal to a given parallelogram, and on the same base. Discuss for impossible cases.

Proposition XIII.

163. Problem. *To construct a square equal to a given polygon.*

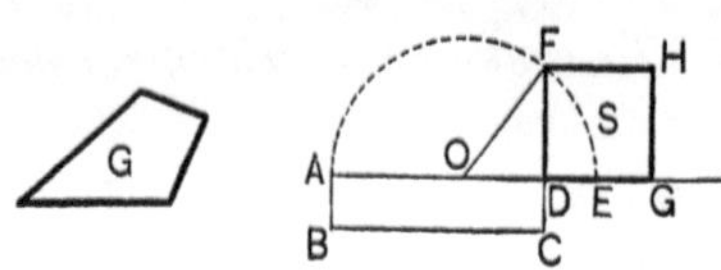

Given   polygon G.

Required   to construct a square equal to G.

Construction.   1. Construct a $\triangle$ equal to G.   Prop. XII

     2. By drawing a line through the vertex of this $\triangle$ ∥ to the base, and erecting ⊥'s from an extremity and the mid-point of the base, construct a $\square$, as $ABCD$, equal to this $\triangle$.   I, props. XXIX, XXXI, XXXIII
Then if $AB = DA$, $ABCD$ is the required $\square$.

     3. If not, produce AD to E, making $DE = CD$;   § 28
bisect AE at O,   I, prop. XXXI
and with center O and radius OE, describe a semi-circumference.   § 109, post. of $\odot$

     4. Produce CD to meet circumference at F,   § 28
and construct a square on DF.   I, prop. XXXIX
Then DF^2, S in the figure, is the required $\square$.

Proof.   1. Draw OF, let $r = OF = OA = OE$, and $x = OD$;
then $CD = DE = r - x,$
and $AD = r + x.$

     2. Then $(r + x)(r - x) = r^2 - x^2$ Prop. VII
$$= x^2 + DF^2 - x^2 = DF^2. \text{ Why?}$$

     3. But $(r + x)(r - x) = \square\, ABCD = G,$ Const. 2
and $\therefore DF^2 = \text{polygon } G.$ Ax. 1

EXERCISES.

215. If one angle of a triangle is two-thirds of a straight angle, show that the square on the opposite side equals the sum of the squares on the other two sides, together with their rectangle.

216. Prove prop. XI for a concave quadrilateral.

217. If $\angle P = 180°$ and $SP = PQ$, show that prop. XI reduces to a previous theorem.

218. Prove prop. XI, cor. directly from prop. X without reference to prop. XI.

219. If $ABCD$ is any quadrilateral, and the mid-points of the diagonals are joined by a line bisected at M, and if P is any point, then $PA^2 + PB^2 + PC^2 + PD^2 = MA^2 + MB^2 + MC^2 + MD^2 + 4\,PM^2$.

220. To construct a parallelogram equal to a given triangle, and having one of its angles equal to a given angle.

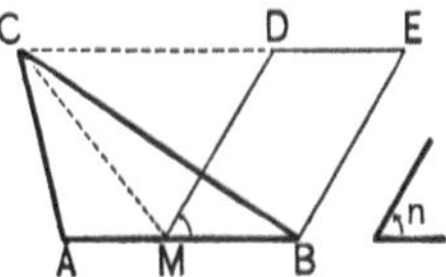

221. To construct a parallelogram equal to a given square, on the same base and having an angle equal to half the angle of the square.

222. To construct an isosceles triangle equal to a given triangle, and on the same base.

223. To construct a triangle equal to a given parallelogram, and having one of its angles equal to a given angle.

224. To construct a parallelogram equal to a given triangle, and having its perimeter equal to that of the triangle. (In the figure of ex. 220 how must MD compare with $BC + CA$?)

225. To construct a square equal to the sum of two given squares. (Apply prop. VIII.)

226. On a given line to construct a rectangle equal to a given rectangle.

227. On one side of a triangle as a diagonal to construct a rhombus equal to the given triangle.

228. Prove that in any triangle three times the sum of the squares on the sides equals four times the sum of the squares on the three medians.

229. Also that three times the sum of the squares on the lines joining the centroid to the vertices equals the sum of the squares on the sides.

230. If one angle of a triangle is one-third of a straight angle, show that the square on the opposite side equals the sum of the squares on the other two sides less their rectangle.

3. PRACTICAL MENSURATION.

164. For practical purposes a surface is measured as follows :

1. A square unit is defined as a square which is one linear unit long and one linear unit wide.

That is, a square inch is a square that is 1 in. long and 1 in. wide; a square meter is a square that is 1 m. long and 1 m. wide, etc. In the figure the shaded square is considered as a square unit.

2. If two sides of a rectangle are 3 in. and 5 in. respectively, then, in the figure, the area of the strip AB is 5×1 sq. in., and the total area is $3 \times 5 \times 1$ sq. in., or 15 sq. in.

Theoretically, a rectangle rarely has sides both of which exactly contain any linear unit, however small. Such cases are discussed in Book IV. But for practical purposes the above method is approximate to any required degree.

At present it is necessary for the student to learn that geometry gives him an instrument for practical work. It will accordingly be assumed that the measurements can be made to any degree of approximation, and that the expressions *area*, *measure*, etc., are understood in their ordinary sense. It has already been explained that the rectangle of two lines corresponds to the product of two numbers; hence, in practice, lines are represented by numbers, and their rectangles by the products of those numbers. This practical measurement will be exemplified hereafter, as it has already been to some extent, in the numerical exercises.

Exercises. 231. A field is in the form of a rhombus, the obtuse angle being twice the acute angle; the shorter diagonal is 300 feet. Find the area of the field in square feet.

232. A railroad embankment extends through a farm 1 mile long, its rails being in straight lines perpendicular to the two parallel sides of the farm; the embankment is 80 ft. wide at the bottom at one end, and 60 ft. at the other. How much land was taken for railroad purposes ?

EXERCISES.

233. A road running across a farm is $\frac{1}{4}$ mile long and 3 rods wide; the road being rectangular, find its area in acres.

234. The side of an equilateral triangle is 15. Find the area.

235. In excavating for a canal 30 ft. deep, 200 ft. wide at the top, and 160 ft. wide at the bottom, what is the area of a cross-section ?

236. One diagonal of a quadrilateral is 100, and the perpendiculars, from the other two vertices, upon it, are 50 and 40. Find the area.

237. The area of a triangle is a and the altitude is h. Find the base. Investigate for $a = 325.85$, $h = 38$; also for $a = 100$, $h = 100$.

238. The area of a trapezoid is a and the two bases are b_1, b_2. Find the altitude. Investigate for $a = 223.375$, $b_1 = 13.5$, $b_2 = 6.4$; also for $a = 10$, $b_1 = 0$, $b_2 = 10$.

239. The area of a trapezoid is 542.5, the altitude is 21.7, and the difference between the parallel sides is 11.2. Find those sides.

240. The area of a square is 2. Find the side of a square of twice the area; thrice the area; four times the area.

241. The altitude of an equilateral triangle is 160. Find the area.

242. The base of an isosceles triangle is $\frac{4}{5}$ of one of the equal sides, and the altitude is 10. Find the area.

243. Two sides of a right-angled triangle are 1036 and 1173. Find the hypotenuse and the area.

244. Find to three decimal places the diagonal of a square whose area is 1.

245. In a right-angled triangle the perpendicular from the vertex of the right angle divides the hypotenuse into two segments, 2.88 and 5.12. Find the two sides.

246. From the vertex A of $\triangle\,ABC$, $AD \perp BC$. Find the lengths of BD, CD, knowing that $AB = 307.8$, $CA = 480.168$, $BC = 689.472$.

247. A surveyor, wishing to erect a perpendicular to a line on the ground, drives two stakes, A, B, 12 links apart ; to these he fastens the ends of a 24-link segment, and stretches the chain, at the end of the 9th link from A, to C. Show that $AC \perp AB$. (This method of erecting perpendiculars was known to the temple and pyramid builders, and surveyors employed for this purpose were called "rope stretchers." The method is still used in practical field work.)

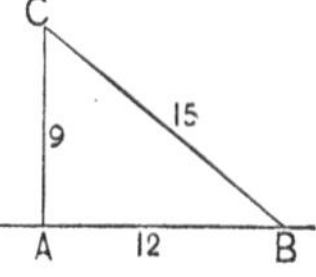

165. **Definitions.** A **circle** is the finite portion of a plane bounded by a curve, which is called the **circumference,** and is such that all points on that line are equidistant from a point within the figure called the **center** of the circle.

For corollaries and postulates, see §§ 108, 109.

Certain definitions are here repeated for convenience.

If two equal figures are necessarily congruent, as in the case of circles, angles, squares, and line-segments, the word *equal* is ordinarily used to express congruence. Hence congruent circles (see § 108, 2) are ordinarily called simply *equal.*

A straight line terminated by the center and the circumference is called a **radius.**

A straight line through the center terminated both ways by the circumference is called a **diameter.**

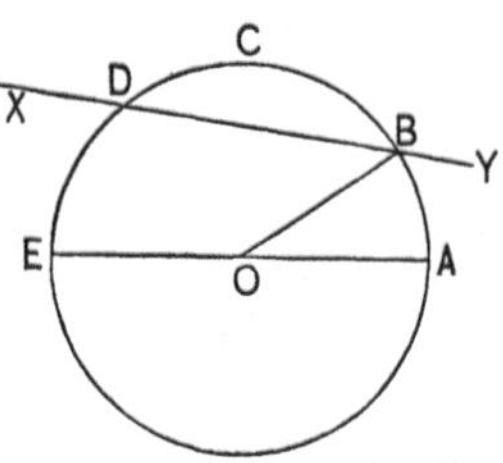

166. The straight line joining any two points on a circumference is called a **chord.**

Hence a diameter is a chord passing through the center. In the figure, *AE* and *BD* are chords.

The expressions center, radius, diameter, chord, of a *circumference* are sometimes used instead of center, etc., of a *circle.*

167. The line of which a chord is a segment is called a **secant,** as *XY* in the figure.

168. A part of a circumference is called an **arc.**

In the figure, *BCD* is an arc. As in naming an angle, the counterclockwise order is followed, and arcs so named are considered positive.

169. One-half of a circumference is called a **semicircumference.**

170. A fourth part of a circumference is called a **quadrant.**

171. An angle formed by two radii is called a **central angle.**

In the figure, $\angle\!s$ AOB, BOE are central angles.

172. A central angle is said to **stand upon** the arc which lies within the angle and is cut off by the arms.

$\angle\!s$ AOB, BOE stand upon $\overset{\frown}{AB}$, $\overset{\frown}{BE}$, respectively.

173. The arc upon which the sum of two central angles stands is called the **sum of the arcs** upon which those angles stand.   Similarly for the **difference of two arcs.**

Thus, $\overset{\frown}{AE} = \overset{\frown}{AB} + \overset{\frown}{BE}$, and $\overset{\frown}{AB} = \overset{\frown}{AD} - \overset{\frown}{BD}$.

174. Two arcs are said to be **complements** of each other if their sum is a quadrant; **supplements** of each other if their sum is a semicircumference; **conjugates** of each other if their sum is a circumference.

In the figure, $\overset{\frown}{AB}$ is the supplement of $\overset{\frown}{BE}$ and the conjugate of $\overset{\frown}{BA}$.

175. An arc greater than a semicircumference is called a **major arc**; one less than a semicircumference, a **minor arc.**

In the figure, AB, BD, DE are minor arcs; DEA is a major arc.

176. Conjugate arcs are said to be **subtended** by their common chord.

In the figure, $\overset{\frown}{BD}$ and $\overset{\frown}{DB}$ are each said to be *subtended* by chord BD.

The word *subtend* is variously used in geometry.   It means to *extend under* or to *be opposite to*.   Hence in a triangle a side is said to subtend an opposite angle, a chord is said to subtend an arc, etc.

177. A portion of a circle cut off by an arc and two radii drawn to its extremities is called a **sector,** and the central angle standing on that arc is called the **angle of the sector.**

In the figure, OAB is a sector, and $\angle AOB$ is its angle.

1. CENTRAL ANGLES.

Proposition I.

178. Theorem. *In the same circle or in equal circles, if two central angles are equal, the arcs on which they stand are equal also, and of two unequal central angles the greater stands on the greater arc.*

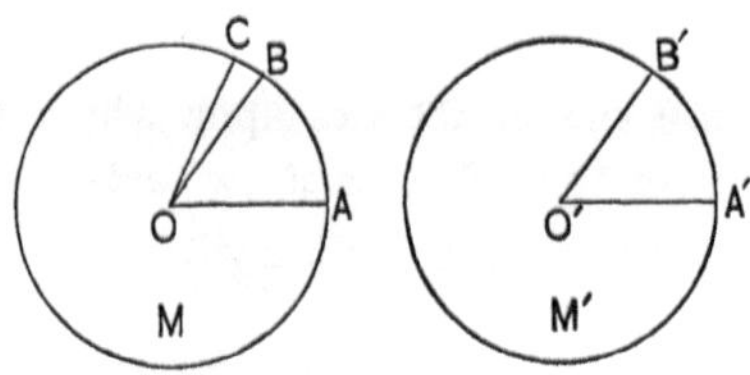

Given M, M', two equal circles, and central angles
$$AOB = A'O'B', \quad AOC > A'O'B'.$$

To prove that $\widehat{AB} = \widehat{A'B'}$, and $\widehat{AC} > \widehat{A'B'}$.

Proof. 1. Place $\odot M'$ on $\odot M$ so that $\angle A'O'B'$ coincides with its equal $\angle AOB$.

Then A' coincides with A, and B' with B.

$\S$ 165, def. $\odot$

2. Then $\widehat{A'B'}$ coincides with $\widehat{AB}$, because its points are equidistant from O. $\S$ 165, def. $\odot$

3. Also, $\because \angle AOC > \angle A'O'B'$,

$\therefore \angle AOC > \angle AOB$.

4. $\therefore C$ is not in $\angle AOB$, and $\widehat{AC} > \widehat{AB}$.   Ax. 8

5. And $\because \widehat{AB} = \widehat{A'B'}$, $\therefore \widehat{AC} > \widehat{A'B'}$.   Ax. 9

The proof is essentially the same for a single circle, and so in general when equal circles are involved.

COROLLARIES. 1. *Sectors of the same circle or of equal circles, which have equal angles, are equal.*

For, by steps 1, 2, they coincide.

2. *Sectors of the same circle, or of equal circles, which have unequal angles, are unequal, the greater having the greater angle.*

This is proved by superposition in steps 3, 4, 5, of the proposition.

3. *The two arcs into which the circumference is divided by a diameter are equal.*

For their central angles are straight angles, and these being equal the arcs are equal by the proposition.

4. *The two figures into which a circle is divided by a diameter are equal.*

For their central angles are straight angles. Hence by cor. 1 they are equal.

This corollary is attributed to Thales.

179. Definition. The figure formed by a semicircumference and the diameter joining its extremities is called a **semicircle.**

It is proved (cor. 4) that all semicircles, cut from the same circle, are equal. Hence the name, *semi-* meaning half.

180. Since the 360 equal angles, into which the perigon at the center of a circle is imagined to be divided, stand on equal arcs by prop. I, the ordinary mensuration of angles by degrees is also used for arcs. Similarly for minutes, seconds, and other measurements. Hence the common expression, *an angle at the center is measured by the subtended arc.*

The expression is not strictly correct ; we do not measure an angle by an arc, but the angle and arc have the same *numerical* measure, as will be proved in § 254. We might as truly say that an arc is measured by its central angle. But the expression is so commonly used, and has found its way into so many text-books and examination papers, that the student needs to become familiar with it.

Proposition II.

181. Theorem. *In the same circle or in equal circles, if two arcs are equal, the central angles which they subtend are equal also, and of two unequal arcs the greater subtends the greater central angle.*

Proof. If O and O' are two central angles, and A, A' are the arcs on which they stand, it has been proved in prop. I that

$$\text{If } O > O', \quad \text{then} \quad A > A',$$
$$\text{`` } O = O', \quad \text{``} \quad A = A',$$
$$\text{`` } O < O', \quad \text{``} \quad A < A'.$$

Hence the converses are true, by the Law of Converse, § 73.

Corollaries. 1. *In the same circle or in equal circles, equal sectors have equal angles; and of two unequal sectors, the greater has the greater angle.*

Law of Converse, § 73, from prop. I, cors. 1, 2.

2. *A central angle is greater than, equal to, or less than, a right angle, according as the arc on which it stands is greater than, equal to, or less than, a quadrant.* (Why?)

Exercises. 248. If two lines drawn to a circumference, from a point within the circle, are equal, they subtend equal central angles.

249. Prove the converse of ex. 248.

250. Two circumferences cannot bisect each other.

251. Suppose from the point P on a circumference two equal chords, PA, PB, are drawn. Prove (1) that these chords subtend equal central angles, (2) that they subtend equal arcs.

252. The arc AB is bisected by the point M, and MC is a diameter; prove that chord $AC = $ chord BC.

253. How many degrees in the central angle standing on a third of a circumference? a fourth? a fifth? .....

2. CHORDS AND TANGENTS.

Proposition III.

182. Theorem. *In the same circle or in equal circles, if two arcs are equal they are subtended by equal chords, and of two unequal minor arcs the greater is subtended by the greater chord.*

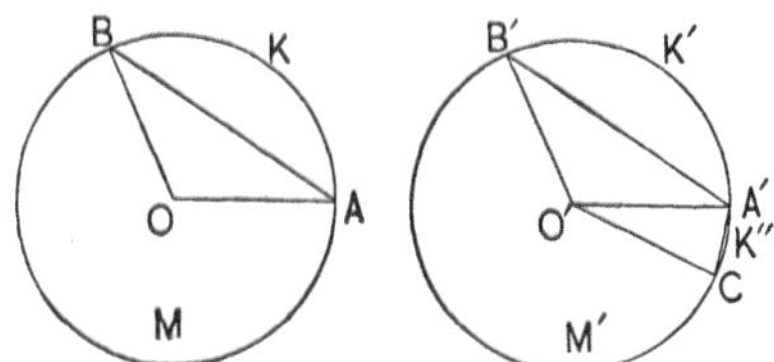

Given   two equal circles, M, M'; two equal arcs, K, K'; and two unequal minor arcs, $K > K''$.

To prove that, as lettered in the figure, chords $AB = A'B'$, $AB > CA'$.

Proof. 1. Draw the radii OA, OB, $O'A'$, $O'B'$, $O'C$.   Then

$$\because \widehat{K} = \widehat{K'}, \quad \therefore \angle AOB = \angle A'O'B'. \qquad \text{Prop. II}$$

2. But $\because \odot M = \odot M'$,

$$\therefore OA = OB = O'A' = O'B' = O'C.$$

3. $\therefore \triangle OAB \cong \triangle O'A'B'$, and $AB = A'B'$.   Why?

4. Also,

$$\because \widehat{K} > \widehat{K''}, \quad \therefore \angle AOB > \angle CO'A', \qquad \text{Prop. II}$$

$$\therefore AB > CA'. \qquad \text{I, prop. X}$$

Corollary. *In the same circle or in equal circles, of two unequal major arcs, the greater is subtended by the less chord.*

Proposition IV.

183. Theorem. *In the same circle or in equal circles, if two chords are equal they subtend equal major and equal minor arcs; and of two unequal chords the greater subtends the greater minor and the less major arc.*

Proof. Let C, C' be two chords of the same circle or of equal circles; N, N' their corresponding minor arcs;

J, J' " " major arcs.

From prop. III,

$$\text{if } N > N', \text{ or if } J < J', \text{ then } C > C',$$
$$\text{" } N = N', \text{ " " } J = J', \text{ " } C = C',$$
$$\text{" } N < N', \text{ " " } J > J', \text{ " } C < C'.$$

Hence the converses are true, by the Law of Converse, § 73.

Exercises. 254. If through a point in a circle two chords are drawn making equal angles with the diameter through that point, these chords cut off equal arcs of the circle.

255. The intersecting chords joining the extremities of two equal arcs of a circle are equal.

256. What is meant by an arc of 75°? by one of 300°? Can the sum of two arcs ever exceed an arc of 360°? Draw a figure to illustrate your answer.

257. Does the chord subtending the arc $2\,a$ equal twice the chord subtending the arc a? Prove your statement.

258. May the chord subtending the arc $2\,a$ ever equal the chord subtending the arc a? If not, show why; if so, tell how many degrees in the arc a.

259. How many degrees in the supplement of the arc 90°? 175°? 180°? 190°?

260. How many degrees in the conjugate of the arc 180°? 300°? 360°? 400°?

261. How does the length of the chord subtending an arc of 60° compare with that subtending an arc of 90°? 300°? (Call the radius r, and determine each in terms of r.)

Proposition V.

184. Theorem. *A diameter which is perpendicular to a chord bisects the chord and its subtended arcs.*

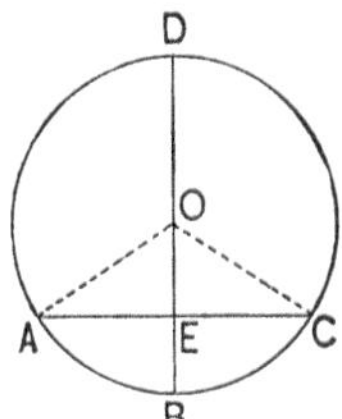

Given   the diameter BD perpendicular to chord AC at E.

To prove that   (1) $AE = EC$,

 (2) $\overset{\frown}{AB} = \overset{\frown}{BC}$,   (3) $\overset{\frown}{DA} = \overset{\frown}{CD}$.

Proof. 1. Drawing radii OA, OC, then

$$OA = OC, \qquad\qquad \text{§ 109, post. of } \odot$$

and $$AE = EC, \qquad\qquad \text{I, prop. XX, cor. 6}$$

$$\therefore \angle AOE = \angle EOC. \qquad \text{I, prop. XX, cor. 5}$$

2. $$\therefore \overset{\frown}{AB} = \overset{\frown}{BC}. \qquad\qquad\qquad \text{Why?}$$

3. And $$\because \angle DOA = \angle COD, \qquad \text{Prel. prop. IV}$$

$$\therefore \overset{\frown}{DA} = \overset{\frown}{CD}. \qquad\qquad\qquad \text{Why?}$$

Corollaries. 1. *Conversely, a diameter which bisects a chord is perpendicular to it.*

For $\because AE = EC$, and $OA = OC$, $\therefore DB$ has two points equidistant from A and C.   Hence, being determined by these points, it is $\perp$ to AC, by I, prop. XLI.

2. *The perpendicular bisector of a chord passes through the center of the circle and bisects the subtended arcs.*

For the center is equidistant from the ends of the chord, by definition of a circle; $\therefore$ it lies on the perpendicular bisector of the chord, by I, prop. XLI.

Proposition VI.

185. Theorem. *All points in a chord lie within the circle ; and all points in the same line, but not in the chord, lie without the circle.*

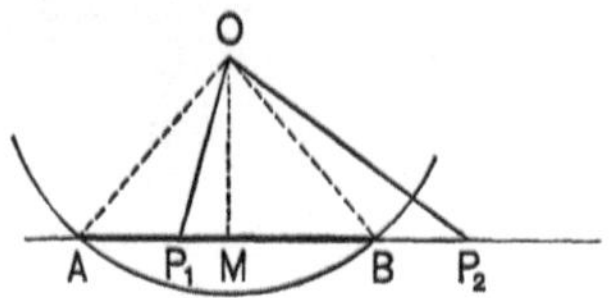

Given the points P_1 in a chord AB, and P_2 in AB produced.

To prove that P_1 is within the circle, and P_2 is without.

Proof. 1. Suppose O the center, and OA, OB, OP_1, OP_2 drawn, and $OM \perp AB$.

Then M is between A and B. 　　　　　　Prop. V

2. And $\because \angle AOM > \angle P_1OM,$

$$\therefore AO > P_1O, \qquad\text{I, prop. XX}$$

and $\therefore P_1$ is within the $\odot$. 　　§ 108, def. $\odot$, cor. 3

3. And $\because \angle MOP_2 > \angle MOB,$

$$\therefore P_2O > BO, \qquad\text{I, prop. XX}$$

and $\therefore P_2$ is without the $\odot$. 　　§ 108, def. $\odot$, cor. 3

Corollary. *A straight line cannot meet a circumference in more than two points.*

For every other point on that line must be either between or not between those two points, and hence must lie either within or without the circle.

Exercises. 262. Prove that, in general, two chords of a circle cannot bisect each other. What is the exception ?

263. What is the locus of the mid-points of a pencil of parallel chords of a circle ? Why ?

Proposition VII.

186. Theorem. *In the same circle or in equal circles, equal chords are equidistant from the center; and of two unequal chords the greater is nearer the center.*

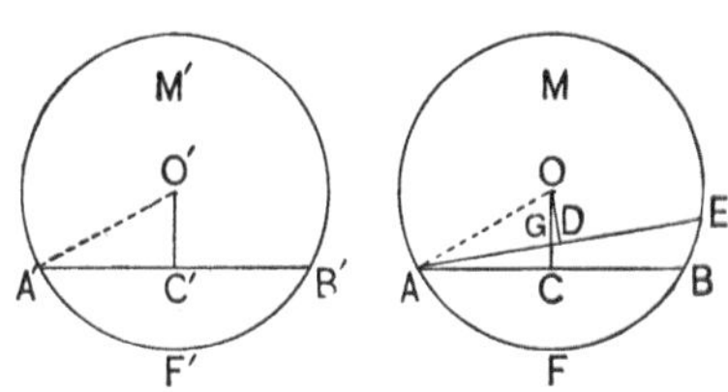

Given two equal ⊙ *M, M'*, with chords $AB = A'B'$, $AE >$
$A'B'$, and OC, OD, $O'C''$ ⊥'s from center O to AB,
AE, and from center O' to $A'B'$.

To prove that (1) $OC = O'C'$, (2) $OD < O'C'$.

Proof. 1. C, C' bisect AB, $A'B'$, Prop. V

$\therefore AC = A'C'$, being halves of equal chords. Ax. 7

2. Draw OA, $O'A'$; then $\because OA = O'A'$,

and $\angle C = \angle C'$, Prel. prop. I

$\therefore \triangle ACO \cong \triangle A'C'O'$, I, prop. XIX, cor. 5

and $OC = O'C'$, which proves (1).

3. And $\because AE > A'B'$,

then $AE > AB$, which equals $A'B'$. Ax. 9

4. $\therefore$ minor $\widehat{AFE} > \widehat{AFB}$,

so that E does not lie on $\widehat{AFB}$. Prop. IV

5. And $\because$ O and AB are on opposite sides of AE,

$\therefore OC$ cuts AE, as at G, and $OD < OG$.

I, prop. XX

6. And $\because OG < OC, \therefore OD < OC.$ Ax. 9

7. And $\because OC = O'C', \therefore OD < O'C'.$ Ax. 9

Proposition VIII.

187. Theorem. *In the same circle or in equal circles, chords that are equidistant from the center are equal; and of two chords unequally distant, the one nearer the center is the greater.*

Proof. If c, c' are two chords of the same circle or of equal circles, and d, d' are the respective perpendiculars from the center upon them; then from prop. VII,

$$\text{If } c > c', \quad \text{then} \quad d < d',$$
$$\text{`` } c = c', \quad \text{``} \quad d = d',$$
$$\text{`` } c < c', \quad \text{``} \quad d > d'.$$

Hence the converses are true by the Law of Converse, § 73.

Corollary. *The diameter is the greatest chord in a circle.* For its distance from the center is zero.

Exercises. 264. AB is a fixed chord of a circle, and XY is any other chord having its mid-point P on AB. What is the greatest and what is the least length that XY can have?

265. What is the locus of the mid-points of equal chords of a circle?

266. Two parallel chords of a circle are 6 inches and 8 inches, respectively, and the distance between them is 1 inch. Find the radius.

267. Two chords are drawn through a point on a circumference so as to make equal angles with the radius drawn to that point. Prove that the chords subtend equal arcs.

268. If from the extremities of any diameter perpendiculars to any secant are drawn, the segments between the feet of the perpendiculars and the circumference will be equal. Draw the various figures.

269. If two equal chords of a circle intersect, the segments of the one are equal respectively to the segments of the other.

270. Find the shortest chord which can be drawn through a given point in a circle.

271. The circumference of a circle whose center lies on the bisector of an angle cuts equal chords, if any, from the arms.

Proposition IX.

188. **Theorem.** *Of all lines passing through a point on a circumference, the perpendicular to the radius drawn to that point is the only one that does not meet the circumference again.*

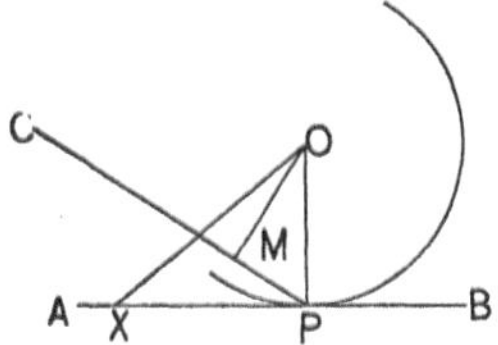

Given   point P on the circumference of a $\odot$ with center O, and AB, PC, respectively perpendicular and oblique to OP at P.

To prove that AB does not meet the circumference again, but that PC does.

Proof   1. Let $OM \perp PC$, and OX be any oblique to AB.

Then $OM < OP$,   I, prop. XX

and $\therefore$ M is within the $\odot$, and PC cuts the circumference again.   §§ 108, 109

2. Also, $OX > OP$,   Why ?

and $\therefore$ X, any point except P on AB, is without the $\odot$.   § 108, def. $\odot$, cor. 3

3. $\therefore$ the perpendicular does not meet the circumference again, but an oblique does.

189. **Definitions.** The unlimited straight line which meets the circumference of a circle in but one point is said to **touch**, or be **tangent** to, the circle at that point. The point is called the **point of contact, or point of tangency,** and the line is called a **tangent.**

A *tangent from a point to a circle* is to be understood as the segment of the tangent between the point and the circle.

If the two points in which a secant cuts a circumference continually approach, the secant approaches the condition of tangency. Hence the tangent is sometimes spoken of as *a secant in its limiting position.*

COROLLARIES. 1. *One, and only one, tangent can be drawn to a circle at a given point on the circumference.*

For the tangent is perpendicular to the radius at that point, and there is only one such perpendicular.   (Has this been proved ?)

2. *Any tangent is perpendicular to the radius drawn to the point of contact.*   (Why ?)

3. *A line perpendicular to a radius at its extremity on the circumference is tangent to the circle.*   (Why ?)

4. *The center of a circle lies on the perpendicular to any tangent at the point of contact.*

For the radius to that point is perpendicular to the tangent, and as there is only one such perpendicular at that point (prel. prop. II), that perpendicular must be the radius.

5. *The perpendicular from the center to a tangent meets it at the point of contact.*

For the radius to that point is perpendicular to the tangent, and there is only one perpendicular from the center to the tangent.

Exercises.   272.   Show that of these three properties of a line, (1) the passing through the center of a circle, (2) the being perpendicular to a given chord, (3) the bisecting of that chord, any two in general necessitate the third.   In what special case is there an exception ?

273.   If a chord is bisected by a second chord, and the second by a third, and the third by a fourth, and so on, the points of bisection approach nearer and nearer the center.

274.   Tangents drawn to a circle from the extremities of a diameter are parallel.

275.   The diameter of a circle bisects all chords which are parallel to the tangent at either extremity.

PROPOSITION X.

190. Theorem. *An unlimited straight line cuts a circumference, touches the circle, or does not meet the circle, according as its distance from the center of the circle is less than, equal to, or greater than, the radius.*

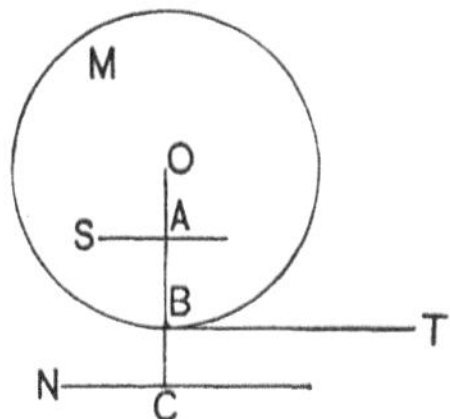

Given OA, OB, OC, the perpendiculars from center O of $\odot M$, to lines S, T, N, and respectively less than, equal to, greater than, the radius.

To prove that S is a secant, T a tangent, N a line not meeting M.

Proof. 1. A, B, C are respectively within the $\odot$, on the circumference, or without the $\odot$.   § 108, def. $\odot$, cor. 3

    2. $\therefore$ S is a secant.   § 109, 2

    3. And T is a tangent.   Prop. IX, cor. 3

    4. $N \parallel T$.   I, prop. XVI, cor. 3

    5. And $\therefore$ N cannot meet $\odot M$ because it cannot cross T.

COROLLARY.   *The converses are true.*

Let the student state this corollary in full, and show that the Law of Converse (§ 73) applies.

———

Exercises.   276. What is the locus of the extremities of equal tangents drawn from points on a circumference ?

277. Two tangents meet at a point the length of a diameter distant from the center of the circle.   How many degrees in their included angle ?

3. ANGLES FORMED BY CHORDS, SECANTS, AND TANGENTS.

191. Definitions. A **segment of a circle** is either of the two portions into which the circle is cut by a chord.

If a segment is not a semicircle, it is called a **major** or a **minor segment** according as its arc is a major or minor arc.

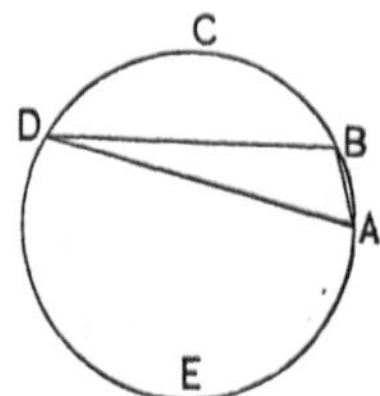

E.g. DBC is a minor segment, and *BDE* is a major segment.

The fact that the word *segment* is used to mean a part of a line, and also a part of a circle, will not present any difficulty, since the latter use is rare, and the sense in which the word is used is always evident. It means "a part cut off," and is therefore applicable to both cases.

192. The angle, not reflex, formed by two chords which meet on the circumference is called an **inscribed angle,** and is said to *stand upon*, or *be subtended by, the arc* which lies within the angle and is cut off by the arms.

It is also called an *angle inscribed in*, or simply an *angle in, the segment* whose arc is the conjugate of the arc on which it stands.

$\angle ADB$ is an inscribed angle, standing on $\overarc{AB}$; it is also an angle in the segment *BCDEA*. Similarly, $\angle DBA$ is in the segment *DAC* and stands on $\overarc{DA}$.

193. Points lying on the same circumference are called **concyclic.**

Exercises. 278. If from the extremities of any chord perpendiculars to that chord are drawn, they will cut off equal segments measured from the extremities of any diameter. (Draw a perpendicular from the center to the chord.)

279. If a tangent from a point *B* on a circumference meets two tangents from *A, C*, on the circumference, in points *X, Y*; and if the lines joining the center to *A, X, Y, C*, are *a, x, y, c*, respectively, then $\angle xy = \angle ax + \angle yc$, and $XY = AX + YC$.

PROPOSITION XI.

194. Theorem. *An inscribed angle equals half the central angle standing on the same arc.*

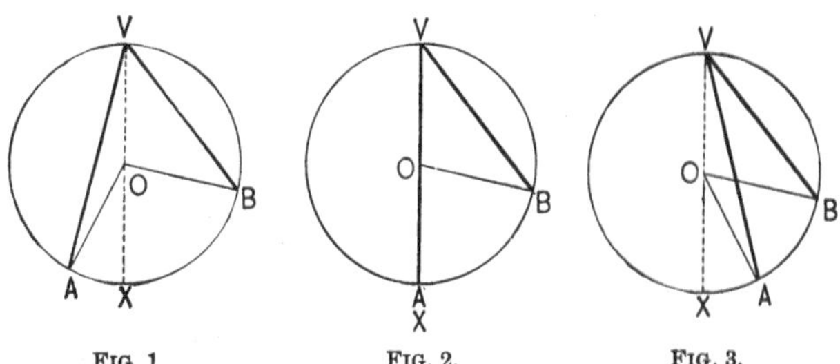

Given AVB an inscribed angle, and AOB the central angle on the same arc AB.

To prove that $\angle AVB = \frac{1}{2} \angle AOB$.

Proof. 1. Suppose VO drawn through center O, and produced to meet the circumference at X.

Then $\angle XVB = \angle VBO$.   I, prop. III

2. And $\angle XOB = \angle XVB + \angle VBO$,   Why?

$= 2 \angle XVB$.   Step 1

3. $\therefore \angle XVB = \frac{1}{2} \angle XOB$.   Ax. 7

4. Similarly $\angle AVX = \frac{1}{2} \angle AOX$ (each $=$ zero in Fig. 2),

and $\therefore \angle AVB = \frac{1}{2} \angle AOB$.   Ax. 2

The proof holds for all three figures, point A having moved to X (Fig. 2), and then through X (Fig. 3).

195. The theorem is often stated thus : *An inscribed angle is measured by half its intercepted arc.*

This expression, like that mentioned in § 180 is not strictly correct. The angle and the arc simply have the same *numerical* measure as proved later in § 254.

Corollaries. 1. *Angles in the same segment, or in equal segments, of a circle are equal.*   (Why?)

2. *If from a point on the same side of a chord as a given segment, lines are drawn to the ends of that chord, the angle included by those lines is greater than, equal to, or less than, an angle in that segment, according as the point is within, on the arc of, or without, the segment.*

This follows from cor. 1 and from I, prop. IX.   Draw the figure and prove.

3. *The converse of cor. 2 is true by the Law of Converse. Hence the locus of the vertex of a constant angle whose arms pass through two fixed points is an arc.*

Let the student state the converse in full, and give the proof.

Exercises. 280.   In the figures on p. 129, prove that if P is taken anywhere on $\overset{\frown}{BV}$, then $\angle PBV + \angle BVP$ is constant.

281. In Fig. 3, p. 129, if BO is produced to meet the circumference at W, and the point of intersection of BW and AV is called Y, prove that $\triangle YVB$ and WAY are mutually equiangular.

282. What is the locus of the vertex of a triangle on a given base and with a given vertical angle?   Prove it.

283. In Fig. 1, p. 129, suppose A to move freely on the arc $VAXB$, and suppose $\triangle AVB$, VBA bisected by lines meeting at P.   Show that the locus of P is a constant arc.

284. If the vertices of a hexagon are concyclic, the sum of any three alternate interior angles is a perigon.   (That is, the sum of three angles, taking every other one.)

285. Two equal chords with a common extremity are symmetric with respect to the diameter through that extremity, as an axis; so also are their corresponding arcs.

286. If from any point P, on the diameter AB, PX and PY are drawn to the circumference on the same side of AB and making $\angle XPA = \angle BPY$, then $\triangle APX$ and YPB are mutually equiangular.

287. If any number of triangles on the same base and on the same side of it have equal vertical angles, the bisectors of the angles are concurrent.

288. Prove that two chords perpendicular to a third chord at its extremities are equal.

Proposition XII.

196. Theorem. *An angle in a segment is greater than, equal to, or less than, a right angle, according as the segment is less than, equal to, or greater than, a semicircle.*

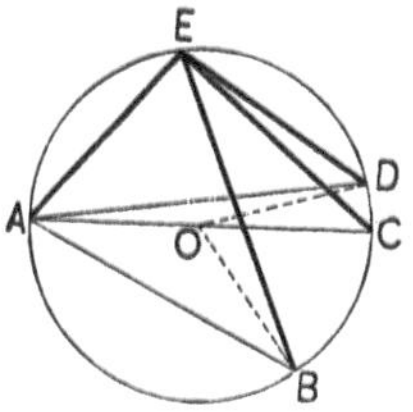

Given   the segments *ADE, ACE, ABE* of a circle with center *O*, respectively less than, equal to, greater than, a semicircle.

To prove   that ⩘ *AED, AEC, AEB* are respectively greater than, equal to, less than, a right angle.

Proof.   1.   Draw *OB, OD.*

Then ∵ $\angle AED = \frac{1}{2}$ reflex $\angle AOD$,   Prop. XI

∴ $\angle AED > $ rt. $\angle$.

2. And   ∵ $\angle AEC = \frac{1}{2}$ st. $\angle AOC$,   Why ?

∴ $\angle AEC = $ rt. $\angle$.

3. And   ∵ $\angle AEB = \frac{1}{2}$ oblique $\angle AOB$,   Why ?

∴ $\angle AEB < $ rt. $\angle$.

Corollary. *A segment is less than, equal to, or greater than, a semicircle, according as the angle in it is greater than, equal to, or less than, a right angle.*

From prop. XII, by the Law of Converse, § 73.   Let the student write out the proof.

Note.   The discovery that an angle in a semicircle is a right angle is attributed to Thales, who, tradition asserts, sacrificed an ox to the gods in honor of the event.

Proposition XIII.

197. Theorem. *An angle formed by a tangent and a chord of a circle equals half of the central angle standing on the intercepted arc.*

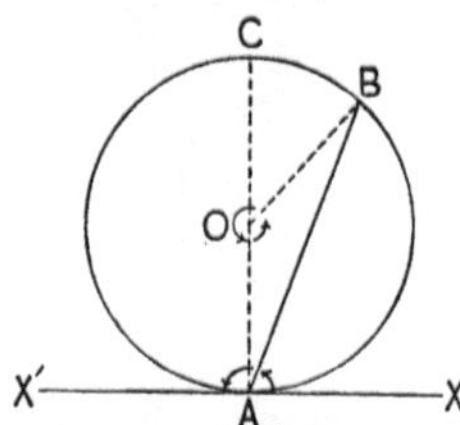

Given AB a chord, XX' a tangent through A, and O the center of the circle.

To prove that $\angle XAB = \tfrac{1}{2} \angle AOB$,

and $\angle BAX' = \tfrac{1}{2} \angle BOA$.

Proof. 1. Produce AO to meet the circumference at C.

Then $\angle XAC = \tfrac{1}{2} \angle AOC$,

$= \tfrac{1}{2}$ st. $\angle$. Why?

2. And $\because \angle BAC = \tfrac{1}{2} \angle BOC$, Why?

$\therefore \angle XAB = \tfrac{1}{2} \angle AOB$. Ax. 3

3. Also, $\because \angle CAX' = \tfrac{1}{2} \angle COA$, Why?

$\therefore \angle BAX' = \tfrac{1}{2} \angle BOA$. Ax. 2

Corollary. *Tangents to a circle from the same external point are equal.*

For, connect the points of tangency, and two angles of the triangle are equal by this theorem.

198. The theorem is often stated thus: *An angle formed by a tangent and a chord of a circle is measured by half its intercepted arc.*

See § 195.

Proposition XIV.

199. Theorem. *An angle formed by two unlimited intersecting lines which meet the circumference equals either the
sum or the difference of half the central angles on the intercepted arcs, according as the point of intersection is within or
without the circle.*

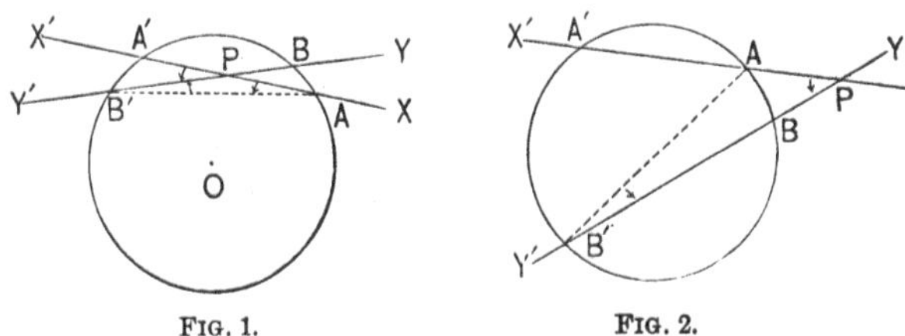

FIG. 1.   FIG. 2.

Given two lines XX', YY' meeting a circumference at
A, A' and B, B', respectively, and intersecting at P.

To prove that $\angle A'PB'$ equals half the central angle on $A'B'$
plus or minus half that on AB, according as P is
within or without the circle.

Proof. Suppose AB' drawn.

Then $\angle A'PB' = \angle A'AB' \qquad \pm \angle AB'B.$ § 88

$= \frac{1}{2}$ cent. $\angle$ on $\overset{\frown}{A'B'} \pm \frac{1}{2}$ cent. $\angle$ on $\overset{\frown}{AB}$.

Prop. XI

The theorem is thus re-stated for two of the special cases :

Corollaries. 1. *An angle formed by two chords equals the
sum of half the central angles on the intercepted arcs.*

See Fig. 1. (State this as suggested in § 195.)

2. *An angle formed by two secants intersecting without the
circle equals the difference of half the central angles on the
intercepted arcs.*

See Fig. 2. (State this as suggested in § 195.)

Prop. XIV is, of course, true for tangents as well as chords and secants. The following figures represent special cases.

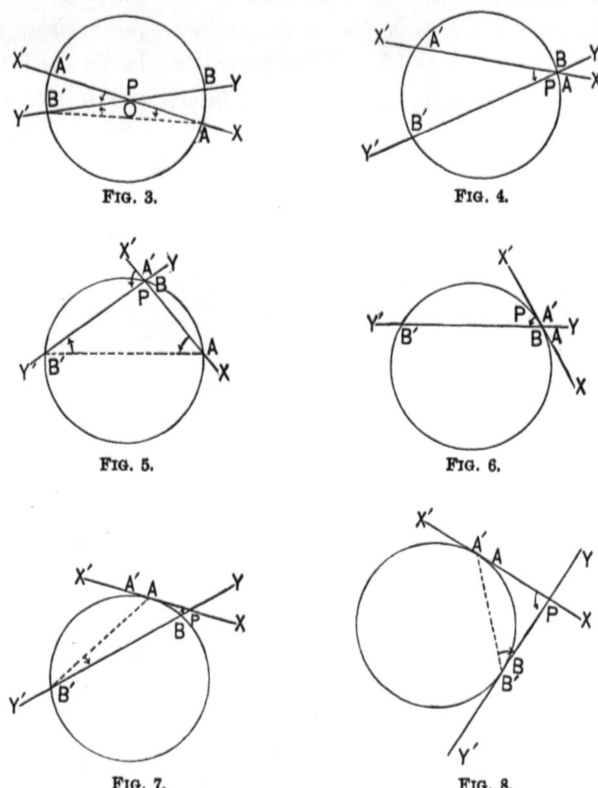

FIG. 3.   FIG. 4.

FIG. 5.   FIG. 6.

FIG. 7.   FIG. 8.

Fig. 3 is a special case where P is at O, and merely affirms that a central angle equals itself. Fig. 4 shows that prop. XI is a special case of prop. XIV. Fig. 6 shows the same for prop. XIII.

COROLLARY. 3. *An angle formed by a secant and a tangent, or by two tangents, equals the difference of half the central angles on the intercepted arcs.*

See Figs. 7 and 8.

PROPOSITION XV.

200. Theorem. *If two parallel lines intercept arcs on a
circumference, those arcs are equal.*

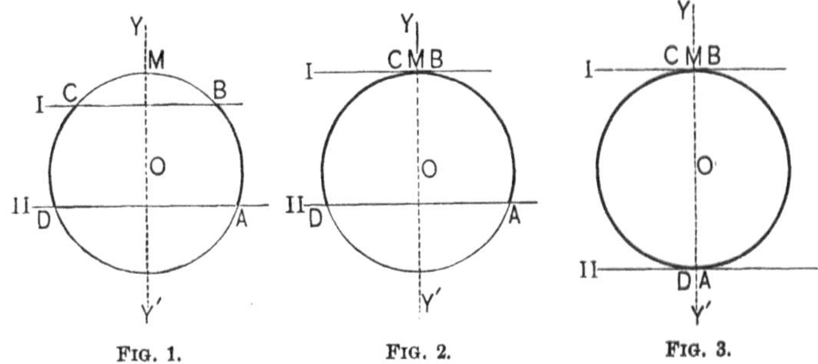

Given   two parallel lines, I and II, intercepting arcs AB,
CD, on the circumference of a circle with center O.

To prove that $\widehat{AB} = \widehat{CD}$.

Proof. 1. Suppose $YOY' \perp I$, and to cut $\widehat{BC}$ at M, Fig. 1.

Then $YY' \perp II$.   I, prop. XVII, cor. 1

2. And $\widehat{BM} = \widehat{MC}$, and $\widehat{AM} = \widehat{MD}$.   Prop. V

3. $\therefore \widehat{AB} = \widehat{CD}$.   Ax. 3

NOTE.   The proof is the same for Figs. 2, 3; in Fig. 2, $\widehat{BC}$ equals zero;
and in Fig. 3, $\widehat{DA}$ also equals zero.   It should be noticed that Figs. 1, 2,
3, respectively, may be considered as special, or at least as limiting cases
of Figs. 2, 7, and 8 of prop. XIV.   In prop. XIV as P moves farther
and farther to the right the lines come nearer and nearer to being parallel,
the angle APB approaches nearer and nearer zero, and hence the central angles on arcs BA, $A'B'$ approach nearer and nearer equality.   It
might therefore be inferred that when the lines become parallel, the arcs
become equal, as proved in prop. XV.

Exercise. 289. The chords which join the extremities of two equal
arcs are either parallel, or else they intersect and are equal and cut off
equal segments from each other.

4. INSCRIBED AND CIRCUMSCRIBED TRIANGLES AND QUADRILATERALS.

201. Definitions. If the vertices of the angles of a polygon lie on a circumference, the polygon is said to be **inscribed in the circle,** and the circle is called a **circumscribed circle.**

If the lines of the sides of a polygon are tangent to a circle, the polygon is said to be **circumscribed about the circle,** and the circle is called an **inscribed** or **escribed circle,** according as it lies within or without the polygon.

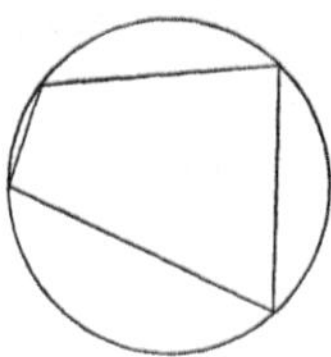

Inscribed quadrilateral.
Circumscribed circle.

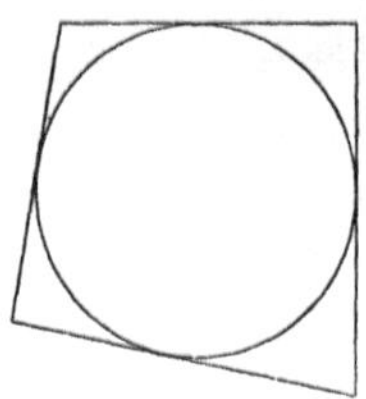

Circumscribed quadrilateral.
Inscribed circle.

Inscribed cross quadrilateral.
Circumscribed circle.

Circumscribed quadrilateral.
Escribed circle.

The words *inscriptible, circumscriptible, escriptible* mean capable of being inscribed in, circumscribed about, escribed to, a circle.

Exercise. 290. If any two chords cut within the circle, at right angles, the sum of the squares on their segments equals the square on the diameter.

Proposition XVI.

202. Theorem. *A circumference can be described to pass through the three vertices of any triangle.* (*Circumscribed circle.*)

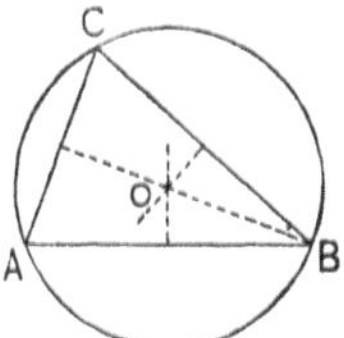

Given   the points A, B, C, the vertices of $\triangle ABC$.

To prove   that a circumference can be described to pass through A, B, C.

Proof. 1. There is such a circumference.   § 131, cor. 2

2. And the center of the $\odot$ can be found.   § 131, cor. 1

Note. The relation between prop. XVI and prop. XVII should be noticed. Similarly for props. XVIII and XIX, and for XX and XXI.

Exercises. 291. Prove from prop. XVI and prop. XI that the sum of the interior angles of any triangle equals a straight angle.

292. If the hypotenuse of a right-angled triangle is the diameter of a circle, the circumference passes through the vertex of the right angle. (Corollary. *The median from the vertex of the right angle of a right-angled triangle equals half of the hypotenuse.*)

293. A line-segment of constant length slides so as to have its extremities constantly resting on two lines perpendicular to each other. Find the locus of its mid-point.

294. If a circle is described on the line joining the orthocenter to any vertex, as a diameter, prove that the circumference passes through the feet of the perpendiculars from the other vertices to the opposite sides.

295. Prove that the perpendiculars from the vertices of a triangle to the opposite sides bisect the angles of the triangle formed by joining their feet; the so-called *Pedal Triangle.*

Proposition XVII.

203. Theorem. *A circle can be described tangent to the three lines of any triangle. (Inscribed and escribed circles.)*

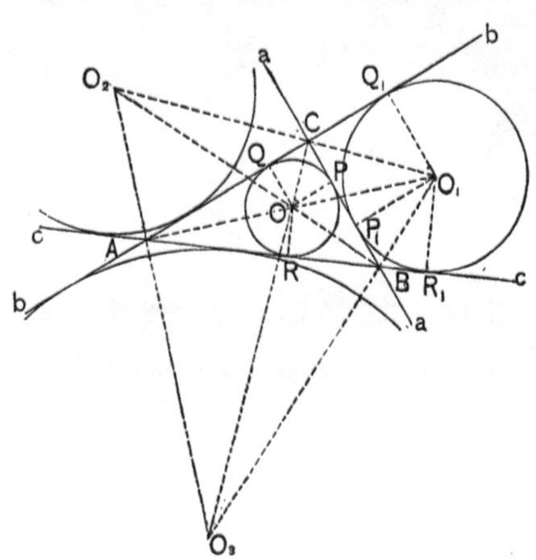

Given   the lines *a*, *b*, *c*, forming a $\triangle ABC$.

To prove   that a circle can be described tangent to *a*, *b*, *c*.

Proof. 1. Let O be the in-center, O_1, O_2, O_3 the ex-centers.

Let OP, OQ, $OR \perp a$, *b*, *c*.

Then $\triangle ARO \cong \triangle AQO$,

and $\triangle BRO \cong \triangle BPO$.   I, prop. XIX, cor. 7

2. $\therefore OQ = OR = OP.$ Why ?

3. $\therefore P, Q, R$ are concyclic.   § 108, def. $\odot$, cor. 3

4. And $\because AB \perp OR$, AB is a tangent.

Prop. IX, cor. 3

Similarly, *a*, *b*, *c*, are tangent to the other three $\circledS$.

Corollary. *A circle can be described tangent to three lines not all parallel nor concurrent.*

Proposition XVIII.

204. Theorem. *In an inscribed quadrilateral the sum or difference of two opposite angles equals the sum or difference of the other two opposite angles, according as the quadrilateral is convex or cross.*

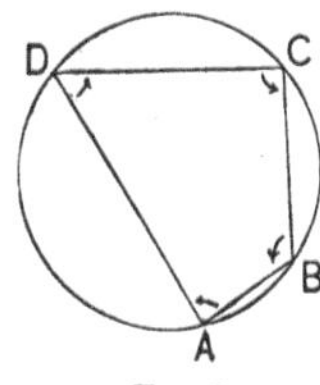

Fig. 1.

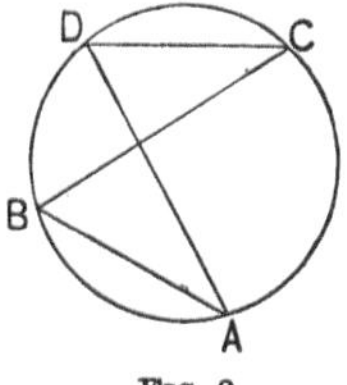

Fig. 2.

Given   the inscribed convex quadrilateral $ABCD$.

To prove   that in Fig. 1, $\angle A + \angle C = \angle B + \angle D$.

Proof for Fig. 1. 1. $\angle A + \angle C = \frac{1}{2}$ central $\angle$ on $\overset{\frown}{BD} + \overset{\frown}{DB}$
$= $ st. $\angle$.   Prop. XI

2. Similarly, $\angle B + \angle D = $ st. $\angle$.

3. $\therefore \angle A + \angle C = \angle B + \angle D$.   § 30

Proof for Fig. 2. If the quadrilateral is cross, $\angle C - \angle A$
$= \angle D - \angle B$, since each equals zero.   Why?

Corollaries. 1. *A parallelogram inscribed in a circle has all of its angles equal, and is therefore a rectangle.*   (Why?)

2. *The opposite angles of an inscribed convex quadrilateral are supplemental.*

Exercises. 296. In the figure of prop. XIII, if P is the mid-point of arc AB, prove that P is equidistant from AX and AB. Suppose the arc BCA is taken, instead of AB.

297. If a circle is described on one side of a triangle as a diameter, prove that the circumference passes through the feet of the perpendiculars drawn to the other two sides from the opposite vertices.

Proposition XIX.

205. Theorem. *In a circumscribed quadrilateral the sum or difference of two opposite sides equals the sum or difference of the other two opposite sides, according as the quadrilateral is convex or cross.*

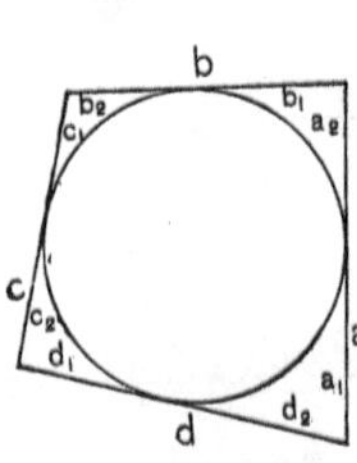

Fig. 1.

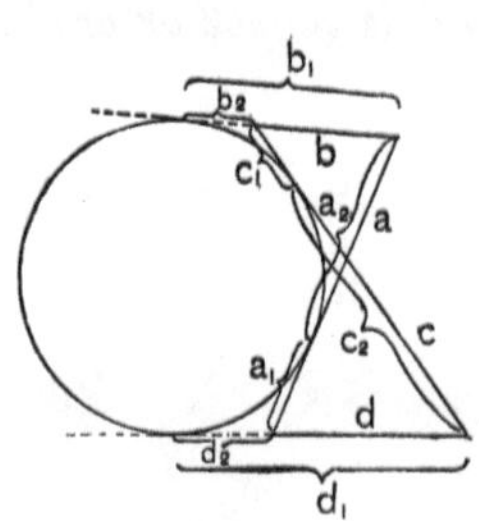

Fig. 2.

Given the circumscribed convex quadrilateral $abcd$.

To prove that in Fig. 1, $a + c = b + d$.

Proof for Fig. 1, as lettered.

    1. $a_1 = d_2,\ \ a_2 = b_1,\ \ c_1 = b_2,\ \ c_2 = d_1.$ Prop. XIII, cor.

    2. $\therefore a_1 + a_2 + c_1 + c_2 = b_1 + b_2 + d_1 + d_2.$ Ax. 2

    3. Or, $a + c = b + d.$ Ax. 8

Proof for Fig. 2. If the quadrilateral is cross, $c - a = d - b$.

    1. $\because c_1 = b_2$, and $c_2 = d_1$, $\therefore c = b_2 + d_1.$

    2. $\because a_1 = d_2$, and $a_2 = b_1$, $\therefore a = b_1 + d_2.$

    3. $\therefore c - a = d - b.$ Ax. 3

Corollary. *A parallelogram circumscribed about a circle has all of its sides equal, and is therefore a rhombus.* (Why?)

Exercises. 298. The bisector of an angle formed by a tangent and chord bisects the intercepted arc.

299. Given two pairs of parallel chords, $AB \parallel A'B'$, and $BC \parallel B'C'$; prove that $AC' \parallel A'C$.

PROPOSITION XX.

206. Theorem. *If the sum of two opposite angles of a quadrilateral equals the sum of the other two opposite angles, the quadrilateral is inscriptible.*

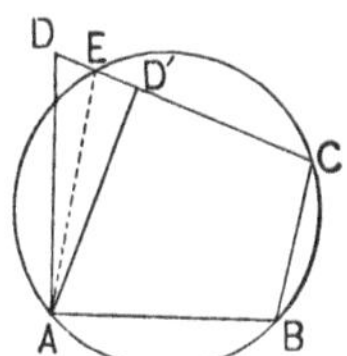

Given   the quadrilateral $ABCD$ such that
$$\angle A + \angle C = \angle B + \angle D.$$

To prove that $ABCD$ is inscriptible.

Proof. 1. Suppose the circumference determined by A, B, C not to pass through D, but to cut CD at E.   Prop. XVI
Draw AE.   Then $\angle B + \angle AEC = \angle C + \angle BAE$.
                                        Prop. XVIII

2. But $\angle B + \angle D = \angle C + \angle A,$

and $\therefore\ \angle AEC - \angle D = \angle BAE - \angle A,$

or, $\angle EAD = - \angle EAD.$ I, prop. XIX

But this is absurd ; hence step 1 is absurd.
The proof is the same for D'.

COROLLARY. *If two opposite angles of a quadrilateral are supplemental, the quadrilateral is inscriptible.*

Exercises. 300. A square is inscriptible.

301. Every equiangular quadrilateral is inscriptible.

302. The intersection of the diagonals of an equiangular quadrilateral is the center of the circumscribed circle.

303. A circle is described on one of the equal sides of an isosceles triangle as a diameter.   Prove that the circumference bisects the base.

PROPOSITION XXI.

207. Theorem. *If the sum of two opposite sides of a quadrilateral equals the sum of the other two opposite sides, the quadrilateral is circumscriptible.*

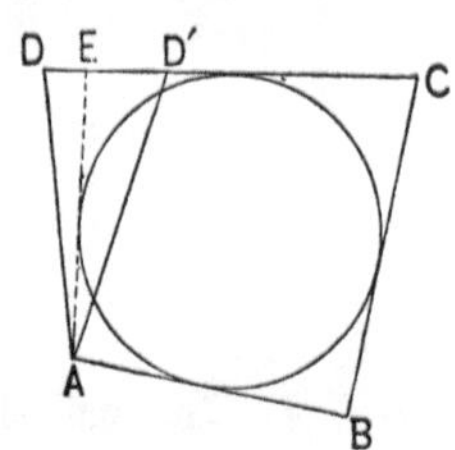

Given　　the quadrilateral $ABCD$ such that
$$AB + CD = BC + DA.$$

To prove　that $ABCD$ is circumscriptible.

Proof. 1. Suppose the $\odot$ tangent to AB, BC, CD not to be tangent to DA, but to be tangent to EA.

Prop. XVII

Then　　　　$AB + CE = BC + EA.$　　　Prop. XIX

2. But　　　　$AB + CD = BC + DA,$　　　　Given

and $\therefore$ $CD - CE$, or ED, $= DA - EA.$　　　Ax. 3

But this is absurd; hence step 1 is absurd.

I, prop. VIII, cor.

The proof is the same for D'.

———

Exercises. 304. A square is circumscriptible. (Notice the relation between exs. 300–302 and exs. 304–306.)

305. Every equilateral quadrilateral is circumscriptible.

306. The intersection of the diagonals of an equilateral quadrilateral is the center of the inscribed circle.

307. A', B' are the feet of perpendiculars from A, B on a, b in $\triangle ABC$; M is the mid-point of AB. Prove that $\angle B'A'M = \angle MB'A' = \angle C$.

5. TWO CIRCLES.

208. Definitions. Two circles are said to **touch** or to be **tangent** when their circumferences have one, and only one, point in common.

They are said to be *internally* or *externally tangent* according as one circle lies within or without the other. The more accurate expression, *a tangent circumference*, is often used instead of *a tangent circle.*

The line determined by the centers of two circles is called their **center-line**; the segment of the center-line, between the centers, is called their **center-segment.**

If two circles have a common center, they are said to be **concentric.**

The expression *concentric circumferences* is also used.

Exercises. 308. A triangle is inscribed in a circle. Prove that the sum of three angles, one in each segment of the circle, exterior to the triangle, equals a perigon.

309. Prove that a perpendicular from the orthocenter of a triangle to a side, produced to the circumference of the circumscribed circle, is bisected by that side.

310. Prove that the bisectors of any angle of an inscribed quadrilateral and the opposite exterior angle meet on the circumference.

311. If the diagonals of an inscribed quadrilateral bisect each other, what kind of a quadrilateral is it ?

312. Prove that if two consecutive sides of a convex hexagon inscribed in a circle are respectively parallel to their opposite sides, the remaining sides are parallel to each other.

313. Prove that the bisectors of the angles formed by producing the opposite sides of an inscribed quadrilateral to meet, are perpendicular to each other. (A proof may be based on cors. 1 and 2 of prop. XIV.)

314. Prove that if the diagonals of an inscribed quadrilateral are perpendicular to each other, the line through their intersection perpendicular to any side bisects the opposite side. (Brahmagupta's theorem.)

Proposition XXII.

209. Theorem. *If two circumferences meet in a point which is not on their center-line, then (1) they meet in one other point, (2) their center-line is the perpendicular bisector of their common chord, (3) their center-segment is greater than the difference and less than the sum of the radii.*

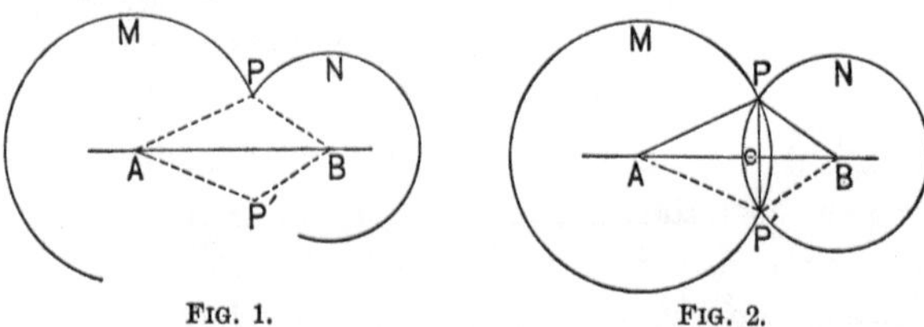

Fig. 1.   Fig. 2.

Given M and N, two circumferences with centers A, B, meeting at P not on AB.

To prove that (1) they meet again, as at P';

(2) $AB \perp PP'$ and bisects it, as at C;

(3) $AB >$ the difference between AP and BP and $< AP + BP$.

Proof. 1. In Fig. 1, suppose $\triangle ABP$ revolved about AB as an axis of symmetry, thus determining $\triangle AP'B$.

Then $\because AP' = AP$, and $BP' = BP$,

$\therefore P'$ is on both M and N, which proves (1).

§ 108, def. ⊙, cor. 3

2. In Fig. 2, $\because AP = AP'$, and $BP = BP'$,   § 109, 1

3. $\therefore A$ and B lie on the $\perp$ bisector of PP', which proves (2).   I, prop. XLI

4. $AB >$ the difference between AP and BP and $< AP + BP$, which proves (3).   § 75 and cor.

Corollary. *If two circumferences meet at one point only, that point is on their center-line.* (Why ?)

Proposition XXIII.

210. Theorem. *If two circles meet on their center-line, they are tangent.*

Given O and O', the centers of two circles with radii OA, $O'A$, which meet on their center-line at A.

To prove that the circles are tangent.

Proof. 1. Let P be any point, other than A, on circumference with center O, and draw OP, $O'P$.

Then $OO' + O'P > OP$ or its equal OA.

I, prop. VIII

2. And $\because OO' = OA - O'A$,

$\therefore OA - O'A + O'P > OA$,

or $O'P > O'A$,

by adding $O'A$ and subtracting OA.   Axs. 4, 5

3. $\therefore P$ is without the circle with center O'.

§ 108, def. ⊙, cor. 3

4. And $\because$ the ⑤ have only one point in common, $\therefore$ they are tangent.   § 208

Corollaries. 1. *If two circumferences intersect, neither point of intersection is on the center-line.* (Why?)

2. *If two circles touch, they have a common tangent-line at the point of contact.*

For a perpendicular to their center-line at that point is tangent to both. (Why?)

—————

Exercise. 315. Find the locus of the centers of all circles tangent to a given circle at a given point.

6. PROBLEMS.

PROPOSITION XXIV.

211. Problem. *To bisect a given arc.*

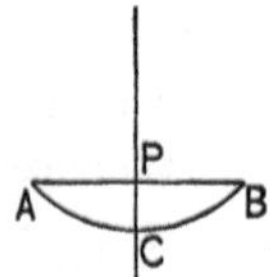

Solution. 1. Draw its chord *AB*.	§ 28
2. Draw $PC \perp AB$ at its mid-point.	§§ 114, 116
Then *PC* bisects the arc.	Prop. V, cor. 2

PROPOSITION XXV.

212. Problem. *To find the center of a circle, given its circumference or any arc.*

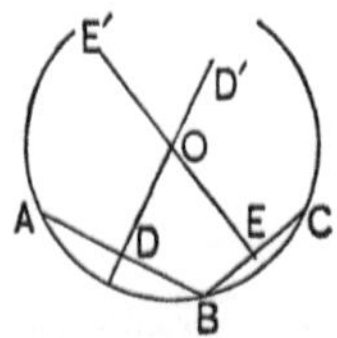

Given a circumference, or an arc *ABC*.

Required to find the center of the circle.

Solution. 1. Draw two chords from *B*, as *BA, BC*.	§ 28
2. Draw their $\perp$ bisectors *DD′, EE′*,	§§ 114, 116
intersecting at the center *O*.	§ 131, cors. 1, 4

NOTE. Hereafter it will be assumed that the center is known if an arc is known, for it may always be found by this problem.

Proposition XXVI.

213. Problem.   *To draw a tangent to a given circle from a given point.*

1. *If the point is on the circumference.*

Solution. 1. At the given point erect a perpendicular to the
radius drawn to the point.   I, prop. XXIX

    2. This is the required tangent, and the solution is
unique.   Prop. IX, cors. 3, 1

2. *If the point is without the circle.*

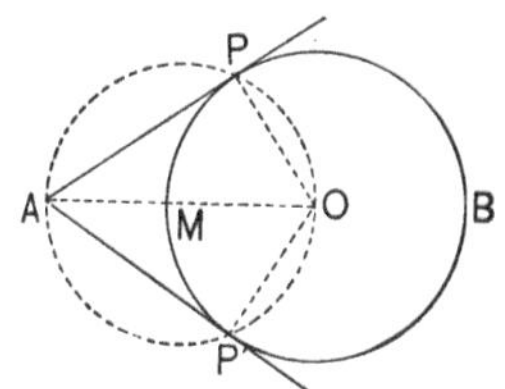

Given   a circle *PP'B*, with center *O*; also an external
point *A*.

Required from *A* to draw a tangent to ⊙ *PP'B*.

Construction. 1. Draw *AO*.   § 28

    2. Bisect *AO* at *M*.   I, prop. XXXI

    3. Describe a ⊙ with center *M*, radius *MO*.   § 109

    4. Join *A* to intersections of circumferences.   § 28
Then these lines, *AP, AP'*, are the required tangents.

Proof. 1. The circumferences will have two points in common,
and only two.   Prop. XXII; I, prop. XLIII, cor. 3

    2. And ∵ ∠ *APO, OP'A* are rt. ∠s,   Why ?
∴ *AP, AP'* are tangents.   Why ?

    (Would this solution hold for case 1 ?)

Proposition XXVII.

214. Problem. *To draw a common tangent to two given circles.*

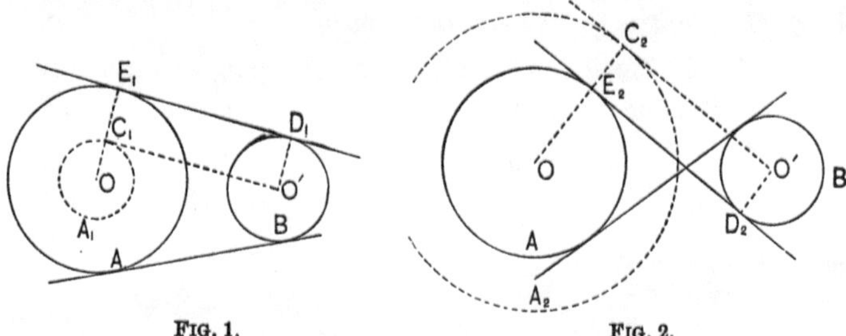

Fɪɢ. 1.   Fɪɢ. 2.

Given   two circles A, B, with radii r, r' $(r > r')$, and centers O, O', respectively.

Required   to draw a common tangent to them.

Construction.   1. Describe ⊚ A_1, A_2 (Figs. 1 and 2), with centers O, and radii $r - r'$ and $r + r'$, respectively.   § 109

     2. From O' draw tangents $O'C_1$, $O'C_2$, to ⊚ A_1, A_2.

Prop. XXVI

     3. Draw OC_1, OC_2, cutting circumferences A at E_1, E_2.

§ 28

     4. Draw $O'D_1 \parallel OE_1$, and $O'D_2 \parallel E_2O$.   § 118

     5. Draw E_1D_1, E_2D_2; they are the tangents.

Proof. 1.   ∡ C_1, C_2 are rt. ∡.   Why?

     2. In Fig. 1,

$$\because C_1E_1 = OE_1 - OC_1 = r - (r - r') = r',$$

$$\therefore C_1E_1 = \text{and} \parallel O'D_1. \qquad \text{Const. 1, 4}$$

     3. $\therefore C_1O'D_1E_1$ is a ▱, and ∡ E_1, D_1 are rt. ∡,

I, props. XXV, XXIII, cor.

and $\therefore D_1E_1$ is tangent to ⊚ A, B.   Prop. IX, cor. 3

Similarly, in Fig. 2, $E_2C_2 =$ and $\parallel D_2O'$, and $E_2D_2O'C_2$ is a $\square$, and D_2E_2 is a tangent.   In both figures a second tangent can evidently be drawn, the solution being analogous to that above given.   Hence there are four tangents in general.

Note.   The following special cases are of interest.

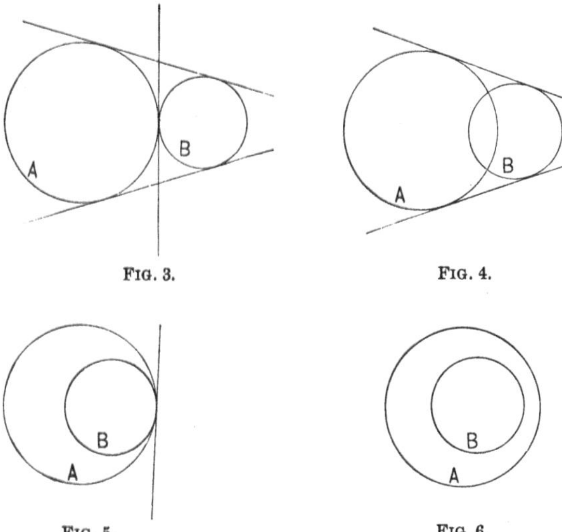

Fig. 3.   Fig. 4.

Fig. 5.   Fig. 6.

In Fig. 3 the two circles have moved to external tangency, and the two interior tangents have closed up into one.   In Fig. 4 the circumferences intersect and the interior tangents have vanished.   In Fig. 5 the circles have become internally tangent and the two exterior tangents have closed up into one.   In Fig. 6 the circle B lies wholly within the circle A, and the tangents have all vanished.   In all cases the center-line is evidently an axis of symmetry.

Exercises.   316. All tangents drawn from points on the outer of two concentric circumferences to the inner are equal.

317. Find the locus of the centers of all circles touching two intersecting lines.   (Show that it is a pair of perpendiculars.)   Suppose the two lines were parallel instead of intersecting.

Proposition XXVIII.

215. Problem. *On a given line-segment as a chord to construct a segment of a circle containing a given angle.*

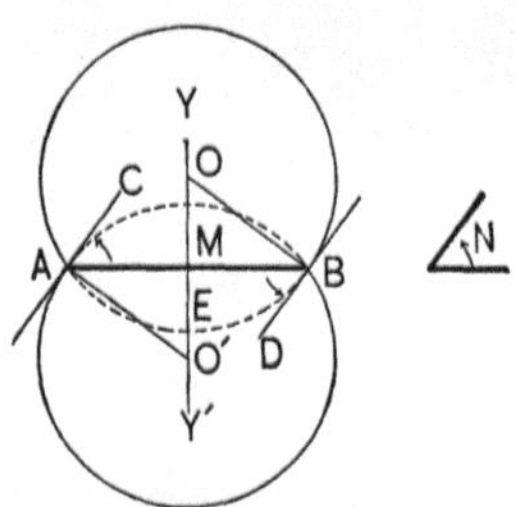

Given the line-segment AB and the $\angle N$.

Required on AB to construct a segment of a circle, containing $\angle N$.

Construction. 1. Draw BD and AC, making $\angle\!\!s$ ABD, BAC equal to $\angle N$. I, prop. XXXII

   2. Draw YY', the $\perp$ bisector of AB. I, prop. XXXI

   3. Draw $\perp$'s to AC, BD, from A, B. I, prop. XXIX

   These $\perp$'s will intersect YY' at the centers of the $\odot$ whose segments on AB are required.

Proof. 1. The two $\perp$'s from A, B, meet YY', as at O', O. I, prop. XVII, cor. 4

   2. O is the center of $\odot$ with chord AB and tangent BD. Prop. V, cor. 2; prop. IX, cor. 4

   3. $\therefore \angle ABD$, or $\angle N$, $= \frac{1}{2}$ central $\angle$ on $A\widehat{E}B$, Prop. XIII

         $= \angle$ in segment ABY. Prop. XI

Similarly for segment $Y'BA$, where $\angle BAC = \frac{1}{2}$ central $\angle$ on the intercepted arc.

216. Definitions. **Two intersecting arcs are said to form an angle,** meaning thereby the angle included by their respective tangents at the point of intersection.

An arc and a secant are said to form an angle, meaning thereby the angle included by the secant and the tangent to the arc at the point of meeting.

E.g. in the figure of prop. XXVIII, *OB* is said to make a right angle with the circumference *EBA*, because it is perpendicular to the tangent at *B*.

Exercises. 318. The bisectors of the interior and the exterior vertical angles of a triangle meet the circumscribed circumference in the mid-points of the arcs into which the base divides that circumference, and the line joining those points is the diameter which bisects the base.

319. A triangle whose angles are, respectively, $30°, 50°, 100°$ is inscribed in a circle; the bisectors of the angles meet the circumference in A, B, C. Find the number of degrees in the angles of $\triangle ABC$.

320. The three sides of $\triangle ABC$ are, respectively, 412 in., 506 in., 514 in.; required the lengths of the six segments formed by the three points of tangency of the inscribed circle.

321. The radii of two concentric circles are 29 in. and 36 in., respectively. In the larger circle a chord is drawn tangent to the smaller; required its length.

322. Two circumferences of circles of radii 0.5 ft. and 1.2 ft. intersect so that the tangents drawn at their point of intersection are perpendicular to each other. Required the distance between the centers.

323. The distance between the centers of two circles of radii 7 in. and 4 in., respectively, is 8 in. Required the length of their common tangent, between the points of tangency. Is there more than one answer?

324. The distance between the centers of two circles of radii 327 in. and 115 in., respectively, is 729 in. Required the length of their common exterior tangent, between the points of tangency.

325. The distance between the centers of two circles is 165 in.; the radii are 62 in. and 48 in., respectively. Calculate, correct to 0.001, the length of the longest line parallel to the center-line and 30 in. from it, limited by the circumferences.

326. Through the point A, 6 in. from the center of a circle of radius 4.5 in., two tangents, AT, AT', are drawn. Calculate the length of the chord TT' and its distance from the center.

APPENDIX TO BOOK III. — METHODS.

217. The student has already been informed of three important methods of attacking a proposition:

(1) By Analysis (§ 113).
(2) By Intersection of Loci (I, props. XLIII, XLIV).
(3) By *Reductio ad Absurdum* (§ 74).

He is now prepared to discuss these somewhat more fully.

218. I. METHOD OF ANALYSIS. This method, first found in Euclid's Geometry, though attributed to Plato, may be thus described: Analysis is a kind of inverted solution; *it assumes the proposition proved, considers what results follow, and continues to trace these results until a known proposition is reached.* It then seeks to *reverse the process* and to give the usual, or *Synthetic*, proof.

A more modern form of analysis is sometimes known as the Method of Successive Substitutions. In this the student substitutes in place of the given proposition another upon which the given one depends, and so on until a familiar one is reached. The student reasons somewhat as follows:

1. I can solve *A* if I can solve *B*.
2. And I can solve *B* if I can solve *C*.
3. But I *can* solve *C*.

Or he reasons thus:

1. *A* is true if *B* is true.
2. And *B* is true if *C* is true.
3. But *C* *is* true.
4. Hence *A* and *B* are true.

Illustrative Exercises.   1. *Through a given point to draw a line to make equal angles with two intersecting lines.*

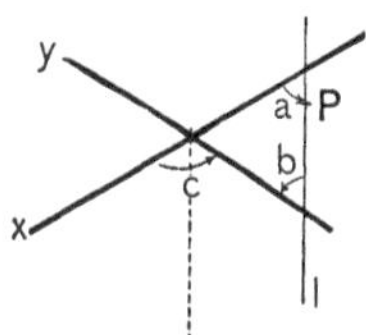

Analysis. Suppose x, y the lines, P the point, and l the required line; then, in the figure, $\angle c = \angle a + \angle b$; but $\because \angle a$ is to equal $\angle b$, $\therefore \angle c = 2 \angle a$; $\therefore$ if $\angle c$ is bisected, and a line is drawn through P parallel to this bisector, the construction is effected. Now that the method is discovered, *give the solution in the ordinary way.*

---

2. *Through a given point to draw a line such that the segments intercepted by the perpendiculars let fall upon it from two given points shall be equal.*

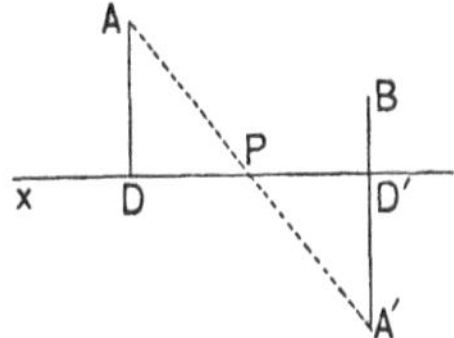

Analysis. Suppose P the given point through which the line x is to be drawn, and A and B the other given points; then, in the figure, AD and $BD' \perp x$, and DP is to equal PD'. Further, if AP is produced to meet BD' produced at A', then $\triangle DPA \cong \triangle D'PA'$, and $\therefore AP = PA'$. But $\because A$ and P are given, AP can be drawn, and PA' found; $\therefore A'$ can be found, and $\therefore A'B$; then from P a $\perp$ can be drawn to $A'B$, and the problem is solved. Always give the solution in the ordinary way.

3. If two circles are tangent, any secant drawn through their point of contact cuts off segments from one that contain angles equal to the angles in the segments of the other.

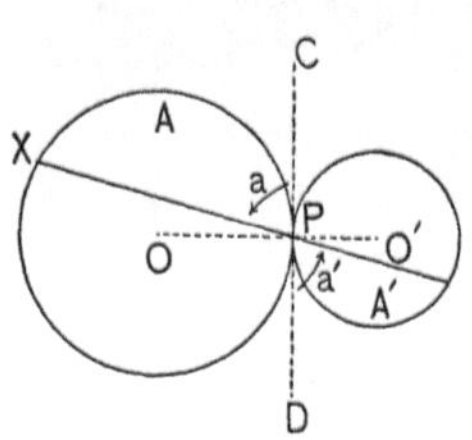

Analysis.   1. Let CD be the common tangent to ⊛ O, O' at their point of contact P.   III, prop. XXIII, cor. 2

2. Then an $\angle$ in segment $A =$ an $\angle$ in segment A', *if*

$$\angle a = \angle a'.$$ III, prop. XIII

3. But $\angle a = \angle a'.$ Prel. prop. VI

Exercises.   327. To construct a trapezoid, given the four sides.

Analysis. Assume the figure drawn. Then if d is moved parallel to itself and between c and a, to the position YZ, the $\triangle XYZ$ can easily be constructed (I, prop. XXXIV). The process may now be reversed and the trapezoid constructed.

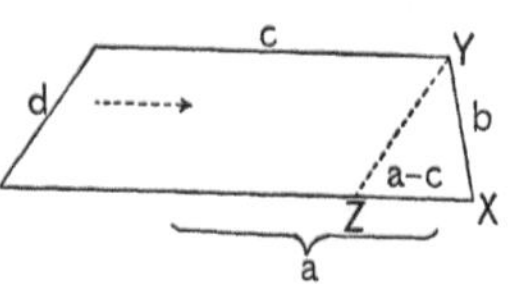

328. To place a line so that its extremities shall rest upon two given circumferences, the line being equal and parallel to another line.

Analysis. If O and O' are the given circles, and AB the given line, and if ⊙ O' is moved along a line parallel and equal to AB, then either XY or $X'Y'$ answers the conditions. Hence the process may be reversed; first describe ⊙ O'', and then from Y, Y' draw YX and $Y'X' =$ and $\parallel BA$.

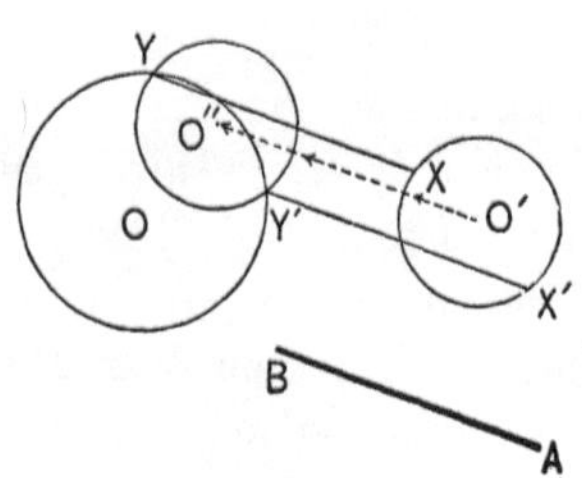

EXERCISES.

329. Given two parallels, XY, $X'Y'$, with a transversal WZ limited by XY and $X'Y'$; also two points A, B, not between the parallels, and on opposite sides of them. Required to join A and B by the shortest broken line which shall have MN, the intercept between XY and $X'Y'$, parallel to WZ.

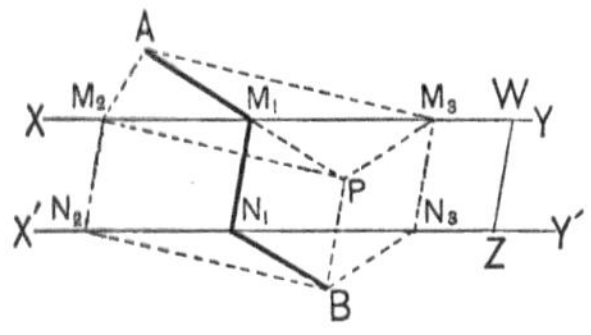

Analysis. If any MN in the figure is moved along NB parallel to its original position, until N coincides with B and M is at P, then $AM_1P <$ AM_2P or AM_3P (I, prop. VIII); hence AM_1N_1B is the shortest broken line. Hence the process may be reversed; first draw $BP \parallel$ and $= ZW$; then join A and P, thus fixing M_1; and then draw $M_1N_1 \parallel WZ$.

330. Through one of the two points of intersection of two circumferences to draw a line from which the two circumferences cut off chords having a given difference. (The projection of the center-segment on the required line equals half the given difference; hence move this projection to the position

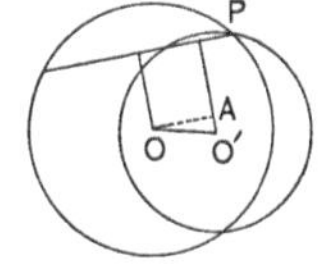

OA; the right-angled $\triangle OO'A$ can now be constructed, and the required line will be parallel to OA.)

331. In ex. 330, show that if the two chords lie on opposite sides of P, the *sum* replaces the *difference*.

332. In a given circle to draw a chord equal and parallel to a given line.

333. From a ship two known points are seen under a given angle; the ship sails a given distance in a given direction, and now the same two points are seen under another known angle. Find the positions of the ship. (On the line joining the known points, construct segments to contain the given angles; the problem then reduces to ex. 328.)

334. Construct a trapezoid, given the diagonals, their included angle, and the sum of two adjacent sides.

335. To construct a triangle given a and the orthocenter.

336. Also, given a and the centroid.

337. To draw a tangent to a given circle, perpendicular to a given line.

338. To construct a triangle, ABC, having given c, $\angle C$, and the foot of the perpendicular from C to c.

339. Find the locus of the points of contact of tangents drawn from a fixed point to a system of concentric circles.

219. II. Method of Intersection of Loci. This method, adapted chiefly to the solution of problems, has already been used in Book I (props. XLIII, XLIV). So long as it is known merely that a point is on *one* line, its position is not definitely known; but if it is known that the point is also on another line, its position *may* be uniquely determined. For example, if it is known that a point is on each of two intersecting lines, the point is uniquely determined as their point of intersection; but if the point is on a straight line and a circumference which the line intersects, it may be either of the two points of intersection.

For convenience of reference the following theorems are stated, and will be referred to by the letters prefixed:

a. The locus of points at a given distance from a given point is the circumference described about that point as a center, with a radius equal to the given distance. (§ 127.)

b. The locus of points at a given distance from a given line consists of a pair of parallels at that distance, one on each side of the fixed line. (§ 129, cor. 2.)

c. The locus of points equidistant from two given points is the perpendicular bisector of the line joining them. (§ 128.)

d. The locus of points equidistant from two given lines consists of the bisectors of their included angles ; if the lines are parallel, it is a parallel midway between them. (§ 129.)

e. The locus of points from which a given line subtends a given angle is an arc subtended by that chord. (§ 195, cor. 3.)

Abbreviations. The following abbreviations will be used: In the triangle ABC the altitudes on the sides a, b, c will be designated by h_a, h_b, h_c, respectively; the corresponding medians by m_a, m_b, m_c; the corresponding angle-bisectors terminated by a, b, c, by v_a, v_b, v_c; the radii of the inscribed and circumscribed circles by r, R, respectively; the radius of the escribed circle touching a, and touching b and c produced, by r_a, and similarly for r_b, r_c.

220. Definition. A *triangle is said to be inscribed in another* when its vertices lie respectively on the sides of the other.

Exercises. 340. To describe a circumference with a given radius, and

(1) Passing through two given points. (Combine a and c.)

(2) Passing through one given point and touching a given line. $(a, b.)$

(3) Passing through one given point and touching a given circle. $(a.)$

(4) Touching a given line and a given circle. $(a, b.)$

(5) Touching two given circles. $(a.)$

341. In a given triangle to inscribe a triangle with two of its sides given, and the vertex of their included angle given. $(a.)$

342. To describe a circumference passing through a given point and touching a given line, or a given circle, in a given point. $(c.)$

343. On a given circumference to find a point having a given distance from a given line. $(b.)$

344. On a given line, not necessarily straight, to find a point equi distant from two given points. $(c.)$

345. Describe a circumference touching two parallel lines and passing through a given point. $(d, a.)$

346. Find a point from which two given line-segments are seen under (or subtend) given angles. $(e.)$ (Pothenot's problem.)

347. Construct the triangle ABC, given a, h_a, m_a.

348. Also, given $\angle A$, a, h_a.

349. Also, given $\angle A$, a, m_a.

350. Also, given a, h_b, h_c.

351. Also, given $\angle A$, h_a, v_a. (First construct the right-angled triangle with side h_a and hypotenuse v_a.)

352. Also, given h_a, m_a, R. (First construct the right-angled triangle with side h_a and hypotenuse m_a; then find the circumcenter by a, c.)

353. Also, given a, R, h_b. (First construct the right-angled triangle with side h_b and hypotenuse a; then find the circumcenter by a.)

354. Also, given c, r, $\angle A = 90°$; c, r_c, $\angle A = 90°$; b, r, $\angle A = 90°$; or b, r_c, $\angle A = 90°$.

355. Describe two circles of given radii r_1, r_2, to touch one another, and to touch a given line on the same side of it.

MISCELLANEOUS EXERCISES.

356. If two circumferences intersect, any two parallel lines drawn through the points of intersection and terminated by the respective circumferences are equal.

357. If the center-segment of two circles is (1) greater than, (2) equal to, (3) less than, the sum of the two radii, the circumferences (1) do not meet, (2) are tangent, (3) intersect.

358. The greatest of all lines joining two points, one on each of two given circumferences, is greater than the center-segment by the sum of the radii.

359. If two circles, whose centers are O, O', are tangent at P, and a line through P cuts the circumferences at A, A', prove that $OA \parallel O'A'$. Two cases; external and internal tangency. Show that the proposition is true for any number of circles.

360. Through a vertex of a triangle to draw a straight line equally distant from the other vertices.

361. Describe a circle of given radius to touch two given lines. Show that a solution is, in general, impossible if the lines are parallel, but that otherwise there are four solutions.

362. From what two points in the plane are two circles seen under equal angles?

363. Given an equilateral triangle, ABC, find a point P such that the circles circumscribing ⧍ PBC, PCA, PAB are all equal.

364. To divide a circle into two segments so that the angle contained in one shall be double that contained in the other.

365. From two given points to draw lines meeting a given line in a point and making equal angles with that line, the points being on (1) the same side of the given line, (2) opposite sides of the given line.

366. To draw, through a given point, a secant from which two equal circumferences shall cut off equal chords. Discuss the number of solutions for various positions of the given point.

367. Through one of the points of intersection of two circumferences to draw a chord of one circle which shall be bisected by the circumference of the other.

368. Two opposite vertices of a given square move on two lines at right angles to each other. Find the locus of the intersection of the diagonals.

369. Find the locus of the intersection of two lines passing through two fixed points on a circumference and intercepting an arc of constant length.

BOOK IV. — RATIO AND PROPORTION.

1. FUNDAMENTAL PROPERTIES.

221. INTRODUCTORY NOTE. The inference was drawn in Book II (§ 155) that a relation exists between algebra and geometry with the following correspondence:

Geometry.	*Algebra.*
A line-segment.	A number.
The rectangle of two line-segments.	The product of two numbers.

And as it was assumed that a straight line may be represented by a number, so it may be assumed that any other geometric magnitude, such as an arc, an angle, a surface, etc., may be represented by a number. With these assumptions, the fundamental properties of Ratio and Proportion may be proved either by algebra or by geometry, as may be most convenient, the proof being valid for both of these subjects. The purely geometric treatment is too difficult for the beginner.

222. Definitions. **To measure a magnitude** is to find how many times it contains another magnitude of the same kind, called the **unit of measure.**

223. A **ratio** is the quotient of the numerical measure of one magnitude divided by the numerical measure of another magnitude of the same kind.

For example, the ratio of a line 8 ft. long to one 16 ft. long is $\frac{8}{16}$, or $\frac{1}{2}$; that of one 16 ft. long to one 8 ft. long is 2.

The ratio of a to b is expressed by the symbols $\frac{a}{b}$, $a : b$, a/b, or $a \div b$.

If the ratio $\frac{a}{b} = r$, then $a = r \cdot b$.

224. *The practical method of finding the ratio* of two magnitudes is to measure them, and to divide the numerical result of one measurement by that of the other. But if two line-segments have a common measure, their ratio and their common measure may be found by the following process:

Let AB and CD be the two lines.

Apply CD as often as possible to AB, and suppose that

$$AB = 2\, CD + EB, \quad EB < CD.$$

Similarly, apply EB to CD, and suppose that

$$CD = 2\, EB + FD, \quad FD < EB.$$

Similarly, apply FD to EB, and suppose that

$$EB = FD + GB, \quad GB < FD.$$

Similarly, apply GB to FD, and suppose that

$$FD = 3\, GB \text{ with no remainder.}$$

Then
$$FD = 3\, GB.$$
$$EB = \quad FD + GB = \quad 4\, GB.$$
$$CD = 2\, EB + FD = \quad 8\, GB + 3\, GB = 11\, GB.$$
$$AB = 2\, CD + EB = 22\, GB + 4\, GB = 26\, GB.$$

$\therefore$ GB is a common measure, and the ratio of CD to AB is $\frac{11}{26}$ by definition of ratio.

225. Definitions. Two magnitudes that have a common measure are said to be **commensurable**; if they have no common measure they are said to be **incommensurable.**

For example, two surfaces having areas 10 sq. in. and 15 sq. in. are said to be commensurable, there being the common measures 5 sq. in., 1 sq. in., 2.5 sq. in., etc. But if the length of one line is represented by $\sqrt{2}$, and the length of another by 1, then there is no common measure, and the lines are said to be incommensurable.

A ratio may therefore be an integer, or a fraction, or an irrational number such as $\sqrt{2}$.

For practical purposes all magnitudes may be looked upon as commensurable, since a unit of measure can be so taken that the remainder may be made as small as we wish.

226. In the ratio $a : b$, a and b are called the **terms** of the ratio, — the former, a, being called the **antecedent,** and the latter, b, the **consequent.**

227. If the ratio $a : b$ equals the ratio $c : d$, the four terms are said to form a **proportion.**

The four terms are also said to be *in proportion.* The terms a and b are also said to be *proportional* to c and d.

This equality of ratios is indicated by the symbol $=$, *e.g.*, $\dfrac{a}{b} = \dfrac{c}{d}$, $a : b = c : d$, or $a/b = c/d$, read "*a* is to *b* as *c* is to *d.*" Instead of the parallel bars ($=$), the double colon ($::$) is also used in this connection as a sign of equality, the proportion being written $a : b :: c : d$. The double colon is not, however, as extensively used as formerly.

228. The first and last terms of a proportion are called the **extremes,** and the other terms the **means.**

Thus in the proportion $2 : 3 = 6 : 9$, 3 and 6 are the means and 2 and 9 are the extremes.

PROPOSITION I.

229. Theorem. *If* $a : b = c : d$, *then* $ad = bc$.

Proof. From $\dfrac{a}{b} = \dfrac{c}{d}$, it follows, by multiplying equals by bd,

that $ad = bc$.   Ax. 6

Hence

If four *numbers* are in proportion, the *product* of the means equals the *product* of the extremes.	If four *lines* are in proportion, the *rectangle* of the means equals the *rectangle* of the extremes.

Proposition II.

230. Theorem. *If* ad $=$ bc, *then* a : b $=$ c : d.

Proof. Divide the given equals by bd. Ax. 7

Hence

If the product of two *numbers* equals the *product* of two other *numbers*, either two may be made the means and the other two the extremes of a proportion.

If the *rectangle* of two *lines* equals the *rectangle* of two other *lines*, either two may be made the means and the other two the extremes of a proportion.

Proposition III.

231. Theorem. *If* a : b $=$ c : d, *then* (1) a : c $=$ b : d, (2) d : b $=$ c : a, *and* (3) b : a $=$ d : c.

Proof. 1.
$$ad = bc,$$ Prop. I

and $$\therefore \frac{a}{c} = \frac{b}{d},$$ which proves (1), Prop. II

and $$\frac{d}{b} = \frac{c}{a},$$ which proves (2). Prop. II

2. And $$\because bc = ad,$$ Step 1

3. $$\therefore \frac{b}{a} = \frac{d}{c},$$ which proves (3). Prop. II

Hence

If four *numbers* are in proportion, the following interchanges may be made : (1) the means, (2) the extremes, (3) each antecedent and its corresponding consequent.

If four *magnitudes* are in proportion, the following interchanges may be made: (1) the means, (2) the extremes, (3) each antecedent and its corresponding consequent.

Definitions. The proportion $a : c = b : d$ is often spoken of as the proportion $a : b = c : d$ *taken by* **alternation.**

The proportion $b : a = d : c$ is also spoken of as the proportion $a : b = c : d$ *taken by* **inversion.**

Hence prop. III may be stated: *If four magnitudes are in proportion they are in proportion by alternation and also by inversion.*

But to take a proportion by alternation, the magnitudes must be similar. Thus \$2 : \$4 = \$8 : \$16, therefore, by alternation, \$2 : \$8 = \$4 : \$16. But the proportion \$2 : \$4 = 8 days : 16 days cannot be taken by alternation, for \$2 : 8 days = \$4 : 16 days means nothing, \$2 : 8 days not being a ratio (§ 223).

———

Proposition IV.

232. **Theorem.** *If* a : b = c : d,

$$\text{then (1) } ka : b = kc : d,$$

$$\text{and (2) } a : kb = c : kd.$$

Proof. From $\dfrac{a}{b} = \dfrac{c}{d}$, it follows, by multiplying by k,

that $\dfrac{ka}{b} = \dfrac{kc}{d}$, which proves (1). Ax. 6

And also, $\dfrac{a}{kb} = \dfrac{c}{kd}$, by dividing by k, which proves (2).

Ax. 7

Hence if four magnitudes are in proportion, and both antecedents or both consequents are multiplied by the same number, the magnitudes are still in proportion.

Corollary. *If four magnitudes are in proportion, and all are multiplied by the same number, the results are in proportion.*

Note. The number k may be integral, fractional, or irrational.

PROPOSITION V.

233. Theorem. *If* $a : b = c : d$,

then   (1) $a \pm b : b = c \pm d : d$,

         (2) $a \pm b : a = c \pm d : c$,

and   (3) $a \pm b : a \mp b = c \pm d : c \mp d$.

Proof. 1. From $\dfrac{a}{b} = \dfrac{c}{d}$, it follows

that $\dfrac{a}{b} \pm 1 = \dfrac{c}{d} \pm 1$,   Axs. 2, 3

or $\dfrac{a \pm b}{b} = \dfrac{c \pm d}{d}$, which proves (1).

2. It is also true

that $\dfrac{b}{a} = \dfrac{d}{c}$,   Prop. III

and $\therefore 1 \pm \dfrac{b}{a} = 1 \pm \dfrac{d}{c}$,   Axs. 2, 3

or $\dfrac{a \pm b}{a} = \dfrac{c \pm d}{c}$, which proves (2).

3. Or, by subtracting first,

$$\frac{a \mp b}{a} = \frac{c \mp d}{c},$$ Axs. 3, 2

and $\therefore \dfrac{a \pm b}{a \mp b} = \dfrac{c \pm d}{c \mp d}$ by dividing in the last two

equations.   Ax. 7

The proportion $a + b : b = c + d : d$ is often spoken of as the proportion $a : b = c : d$ *taken by* **composition**, and $a - b : b = c - d : d$ as the same proportion *taken by* **division**, and $a \pm b : a \mp b = c \pm d : c \mp d$ as the same proportion *taken by* **composition and division.**

Proposition VI.

234. Theorem. *If $\dfrac{a_1}{b_1} = \dfrac{a_2}{b_2} = \dfrac{a_3}{b_3} = \ldots\ldots$, the terms all being magnitudes of the same kind, then*

$$\frac{a_1 + a_2 + \ldots\ldots}{b_1 + b_2 + \ldots\ldots} = \frac{a_1}{b_1} \text{ or } \frac{a_2}{b_2} \text{ or } \ldots\ldots$$

Proof. 1.

$$\frac{a_1}{a_2} = \frac{b_1}{b_2},$$

Prop. III

and

$$\therefore \frac{a_1 + a_2}{a_2} = \frac{b_1 + b_2}{b_2}.$$

Prop. V

2.

$$\therefore \frac{a_1 + a_2}{b_1 + b_2} = \frac{a_2}{b_2}$$

Prop. III

$$= \frac{a_3}{b_3} = \ldots\ldots$$

Given

3. Then, as in steps 1, 2,

$$\frac{a_1 + a_2 + a_3}{b_1 + b_2 + b_3} = \frac{a_3}{b_3} = \ldots\ldots$$

and so on, however many ratios there may be.

Proposition VII.

235. Theorem. *If* $a : b = c : d$, *then* $a^2 : b^2 = c^2 : d^2$.

Proof.

$$\because \frac{a}{b} = \frac{c}{d}, \quad \therefore \frac{a^2}{b^2} = \frac{c^2}{d^2}.$$

Ax. 6

Hence

If four *numbers* are in proportion, *their squares* are also in proportion.	If four *lines* are in proportion, the *squares on those lines* are also in proportion.

Corollary. *If* $a : b = c : d$, *and* $m : n = x : y$, *then* $am : bn = cx : dy$.

Proposition VIII.

236. Theorem. $a : b = ka : kb.$

Proof. $\qquad kab \equiv kab,$ or $a \cdot kb = b \cdot ka.$

$$\therefore a : b = ka : kb. \qquad\qquad \text{Prop. II}$$

Note. The number k may be integral, fractional, or irrational.

237. Definitions. If $a : b = c : x$, x is called the **fourth proportional** to a, b, c.

Corollaries. 1. *By three of the four terms of a proportion the other is determined.*

For if $a : b = c : x$, or $x : b = c : a$, or $c : x = a : b$, etc., it follows that $ax = bc$, whence $x = bc/a$, a fixed number.

2. *If* $a : b = a : x$, *then* $b = x.$

For if, in the proof of cor. 1, $c = a$, then $b = x.$

238. If, in a proportion, the two means are equal, as in $a : x = x : b$, this common mean is called the **mean proportional,** or **geometric mean,** between the two extremes.

Corollaries.

The mean proportional *between two* numbers *equals the* square root *of their* product.	*The* geometric mean *between two* lines *equals the* side of that square *which equals their* rectangle.

Because the number representing the square units of area of a rectangle is the product of the two numbers representing the linear units in two adjacent sides, the expression *product of two lines* is often used for *rectangle of two lines.*

Exercises. **370.** Find a mean proportional between 2 and 32.

371. Find a fourth proportional to 3, 7, 15.

372. What number must be added to each of the numbers 2, 1, 5, 3, to have the results in proportion ?

2. THE THEORY OF LIMITS.

239. **Definitions.** A quantity is called a **variable** if, in the course of the same investigation, it may take indefinitely many values; on the other hand, a quantity is called a **constant** if, in the course of the same investigation, it keeps the same value.

$E.g.$ if a line AB is bisected at M_1, and M_1B at M_2, and M_2B at M_3, and so on, and if x represents the line from A to any of the points M_1, M_2, ..., then x is a variable, but AB is a constant.

It is customary, as in algebra, to represent variables by the last letters of the alphabet, and constants by the first letters.

240. If a variable x approaches nearer and nearer a constant a, so that the difference between x and a can become and remain smaller than any quantity that may be assigned, then a is called the **limit** of x.

$E.g.$ in the above figure, AB is the limit of x.

But if the point M simply slides along the line, passing through B, then, although the difference between AM and AB, or x and a, can *become* smaller than any quantity which may be assigned, it does not *remain* smaller, for when M passes through B this difference increases. Hence AB is not then the limit of x.

That "x approaches as its limit a" is indicated by the symbol $x \doteq a$.

CoROLLARY. *If* $x \doteq a$, *then* $a - x$ *is a variable whose limit is zero ; that is,* $a - x \doteq 0$.

Although the variable has been taken, in this discussion, as *increasing* towards its limit, it may also be taken as *decreasing*. Thus if we bisect a line, bisect its half, and continue to bisect indefinitely, the variable segment is evidently approaching a limit zero.

Proposition IX.

241. Theorem. *If, while approaching their respective limits, two variables have a constant ratio, their limits have that same ratio.*

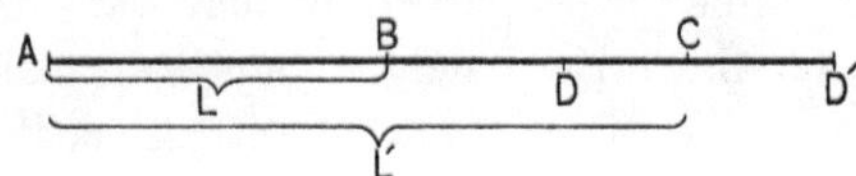

Given X and X', two variables, such that as they *increase* they approach their respective limits AB, or L, and AC, or L', and have a constant ratio r.

To prove that $L : L' = r$, or that $X : X' = L : L'$.

Proof.   If the ratio $X : X'$ is not equal to the ratio $L : L'$, then (1) it must equal the ratio of L to something less than L', or (2) it must equal the ratio of L to something greater than L'.

It will be shown that both of these suppositions are absurd.

I. *To show that* (1) *is absurd.*

1. Suppose $X : X' = L : L' - DC.$
   Then $\because X : X' = r,$
               $\therefore X = rX',$
   and $L = r(L' - DC).$ § 223, def. ratio

2. Then $L - X = r(L' - DC - X').$ Ax. 3

3. But $\because X' \doteq L',$

   $\therefore L' - X'$ can become as small as we please.

   $\therefore$ "   "   "   less than $DC,$

   and $\therefore r(L' - X' - DC)$ can become negative.

4. But $\because X \not> L$, $\therefore L - X$ cannot become negative.

5. $\therefore$ step 2 is absurd, for a negative quantity cannot equal one not negative.

II. *To show that* (2) *is absurd.*

    1. Suppose $X : X' = L : L' + CD'.$

      Then $L - X = r(L' + CD' - X'),$

    as in step 2, p. 168.

    2. But $r(L' + CD' - X')$ cannot become less than $r \cdot CD'.$

    3. And $L - X \doteq 0$, because L is the limit of X.

    4. $\therefore$ if step 1 were true, a quantity, $L - X$, which can become as small as we please, would equal a quantity not less than $r \cdot CD'$, which is absurd.

The proof would be substantially the same if the two variables were supposed to *decrease* toward a limit.

Corollaries. 1. *If, while approaching their respective limits, two variables are always equal, their limits are equal.*

For their ratio is always 1.

2. *If, while approaching their respective limits, two variables have a constant ratio, and one of them is always greater than the other, the limit of the first is greater than the limit of the second.*

For the limits have the same ratio as the variables.

Exercises. **373.** If $a : b = c : d,$

prove that $a^2bd + b^2c + bc = ab^2c + abd + ad.$

    1. Since $bc = ad,$ Prop. I

the equation is true if $a^2bd + b^2c = ab^2c + abd.$

    2. Now if in place of each ad we put bc, we see that the equation is true if $ab^2c + b^2c = ab^2c + b^2c.$

But this is an identity. Hence the proof is complete.

    374. If $a : b = c : d,$

prove that $a - c : b - d = \sqrt{a^2 + c^2} : \sqrt{b^2 + d^2}.$

Also that $\sqrt{a^2 + c^2} : \sqrt{b^2 + d^2} = \sqrt{ac + \dfrac{c^3}{a}} : \sqrt{bd + \dfrac{d^3}{b}}.$

Also that $a + mb \cdot a - nb = c + md : c - nd.$

3. A PENCIL OF LINES CUT BY PARALLELS.

242. Definitions. Through a point any number of lines can be passed. Such lines are said to form a **pencil of lines.**

The point through which a pencil of lines passes is called the **vertex** of the pencil.

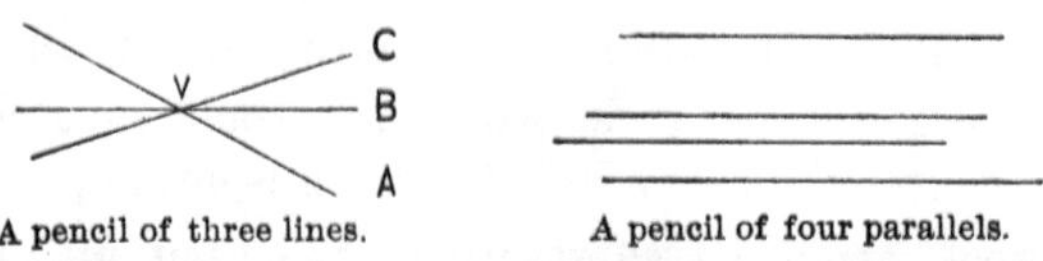

A pencil of three lines.   A pencil of four parallels.

The annexed pencil of three lines is named "$V - ABC$."

To conform to the idea of a general figure, set forth in §§ 94, 95, the word *pencil* is also applied to parallel lines, the vertex being spoken of as "at infinity."

243. Definition. Two lines are said to be **divided proportionally** when the segments of the one have the same ratio as the corresponding segments of the other.

Exercises. **375.** If $a : b = c : d$, prove that

(1) $a + b + c + d : b + d = c + d : d.$

(2) $m (a + mb) : n (a - nb) = m (c + md) : n (c - nd).$

(3) $a (a + b + c + d) = (a + b) (a + c).$

(4) $a^2c + ac^2 : b^2d + bd^2 = (a + c)^3 : (b + d)^3.$

376. If b is a mean proportional between a and c, prove that

$$\frac{a^2 - b^2 + c^2}{\dfrac{1}{a^2} - \dfrac{1}{b^2} + \dfrac{1}{c^2}} = b^4.$$

377. Show that there is no finite number which, when added to each of four unequal numbers in proportion, will make the resulting sums in proportion.

378. If $a : b = c : d$, and $u : v = x : y$,

prove that $au + bv : au - bv = cx + dy : cx - dy.$

PROPOSITION X.

244. Theorem. *The segments of a transversal of a pencil of parallels are proportional to the corresponding segments of any other transversal of the same pencil.*

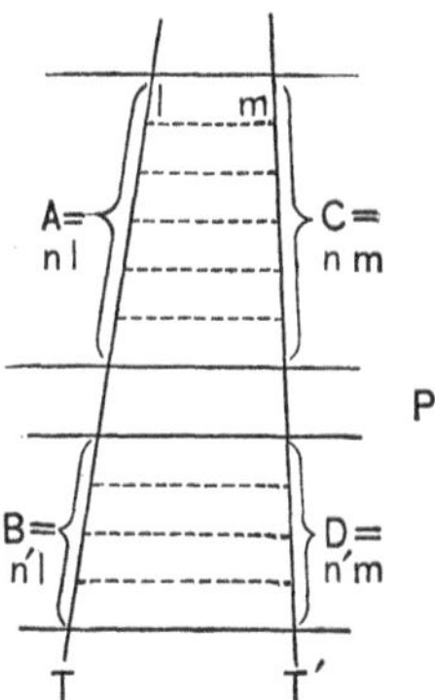

Given   the pencil of parallels P, cutting from two transversals T and T' the segments A, B and C, D, respectively.

To prove that $A : B = C : D.$

Proof. 1. Suppose A and B divided into equal segments l, and that $A = nl$, while $B = n'l$.

(In the figure, $n = 6$, $n' = 4$.)

Then if $\parallel$'s to P are drawn from the points of division, C is the sum of n equal segments m, and D is the sum of n' equal segments m.

I, prop. XXVII, cor. 1

2. $\therefore \dfrac{A}{B} \equiv \dfrac{nl}{n'l} = \dfrac{n}{n'} = \dfrac{nm}{n'm} \equiv \dfrac{C}{D}.$ Why ?

NOTE. The preceding proof assumes that A and B are commensurable. The following proof is valid if A and B are incommensurable.

245. Proof for incommensurable case.

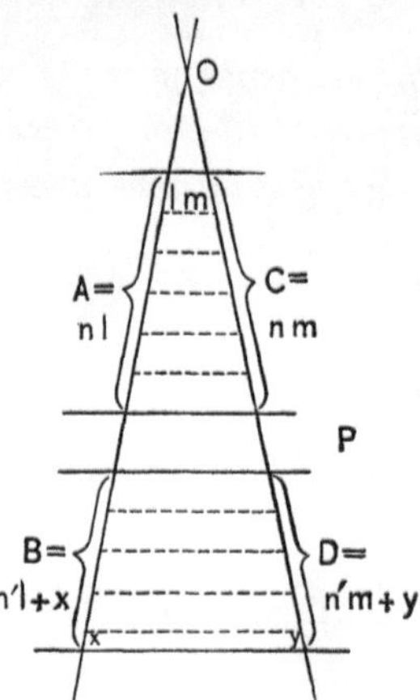

1. Suppose A divided into equal segments l, and that

$$A = nl,$$

while $B = n'l +$ some remainder, x, such that $x < l$. Then if ‖'s to P are drawn from the points of division, C is the sum of n equal segments m, and D is the sum of n' equal segments m, $+$ a remainder y such that $y < m$.

2. Then B lies between $n'l$ and $(n'+1)\,l$. Step 1

3. $\therefore \dfrac{B}{A}$ lies between $\dfrac{n'l}{nl}$ and $\dfrac{(n'+1)\,l}{nl}$

  $\left(\text{in the figure, between } \dfrac{4l}{6l} \text{ and } \dfrac{5l}{6l}\right),$

  while $\dfrac{D}{C}$ lies between $\dfrac{n'm}{nm}$ and $\dfrac{(n'+1)\,m}{nm}$.

4. $\therefore \dfrac{B}{A}$ and $\dfrac{D}{C}$ both lie between $\dfrac{n'}{n}$ and $\dfrac{n'+1}{n}$,

  and $\therefore$ they differ by less than $\dfrac{1}{n}$.

   (In the figure, by less than $\tfrac{1}{6}$.)

5. And $\because \dfrac{1}{n}$ can be made smaller than any assumed difference, by increasing n, $\therefore$ to assume any difference leads to an absurdity.

6. $$\therefore \frac{B}{A} = \frac{D}{C}, \text{ or } \frac{A}{B} = \frac{C}{D}.$$ Prop. III

COROLLARIES. 1. *A line parallel to one side of a triangle divides the other two sides proportionally.*

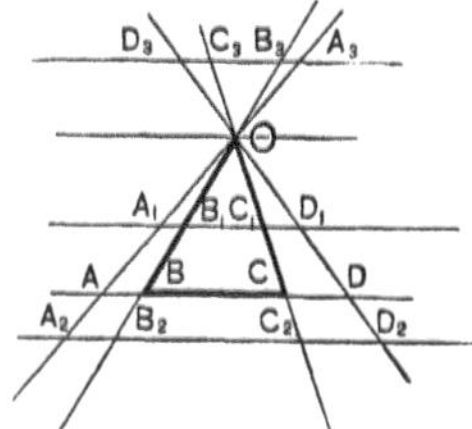

For in the figure, if BCO is the triangle, the lines OB, OC are cut by parallels. Hence $BB_1 : B_1O = CC_1 : C_1O$.

2. *The corresponding segments of the lines of a pencil cut off (from the vertex) by parallel transversals are proportional.*

In the above figure, $OA : OA_1 = OB : OB_1 = OC : OC_1 = \ldots\ldots$, by prop. X.

3. *The segments of the lines of a pencil cut off (from the vertex) by parallel transversals are proportional to the corresponding segments of the transversals.*

To prove that, in the above figure, $AB : A_1B_1 = OA : OA_1 = OB : OB_1$. Draw through A_1 a line $\parallel$ to OB cutting AB at X.

Then　　　　$AB : XB = OA : OA_1 = OB : OB_1$.　　　　Prop. X

But　　　　$XB = A_1B_1$.　　　　I, prop. XXIV

4. *Parallel transversals are divided proportionally by the lines of a pencil.*

To prove that, in the above figure, $AB : BC = A_1B_1 : B_1C_1$.
By cor. 3, $AB : A_1B_1 = BO : B_1O = BC : B_1C_1$.　　Hence, etc.

Proposition XI.

246. **Theorem.** *A line can be divided, internally or externally, into segments having a given ratio, except that if it is divided externally the ratio cannot be unity.*

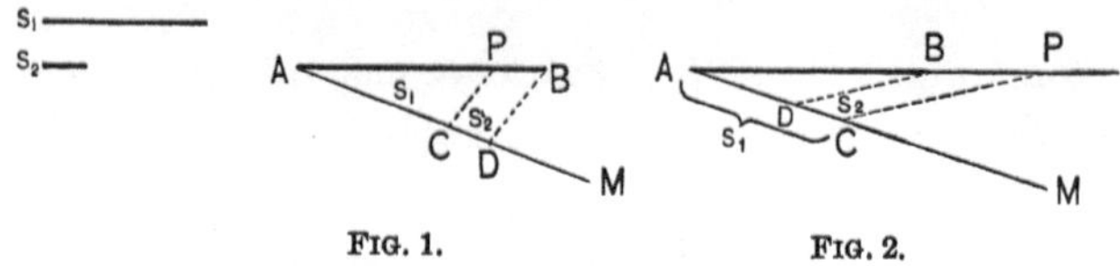

FIG. 1.　　　　　　　　FIG. 2.

Given　　　the line AB, and two lines s_1, s_2 having a given ratio.

To prove　that AB can be divided in the ratio $s_1 : s_2$, except that in the case of external division s_1 cannot equal s_2.

Proof.　1.　Suppose AM drawn making, with AB, an angle $< 180°$; that AC be taken $= s_1$, and $CD = s_2$; that DB be drawn, and $CP \parallel DB$.

　　　　2.　Then $AP : PB = s_1 : s_2$, as required.　　Prop. X, cor. 1

　　　　3.　In Fig. 2, if $s_1 = s_2$, where does D fall ?　What is then the relation of CP to AB ?　Hence show that the division is impossible in this case.

Corollary.　*The point of internal division is unique; likewise the point of external division.*

From step 2, $AB : PB = s_1 + s_2 : s_2$, AB, $s_1 + s_2$, and s_2, all being constants; but by three terms of a proportion the fourth is determined. (§ 237, def. of 4th proportional, cor. 1.)

247.　Note.　Instead of saying that the external division, if the ratio is unity, is impossible, it is often said that the point of division, P, is *at infinity.*

In the case of internal division, the ratios $AP : PB$ and $AC : CD$ are evidently positive ; but in the case of external division each ratio is evidently negative because PB and CD are negative.　In both cases step 2 is evidently true.

Proposition XII.

248. Theorem. *A line which divides two sides of a triangle proportionally is parallel to the third.*

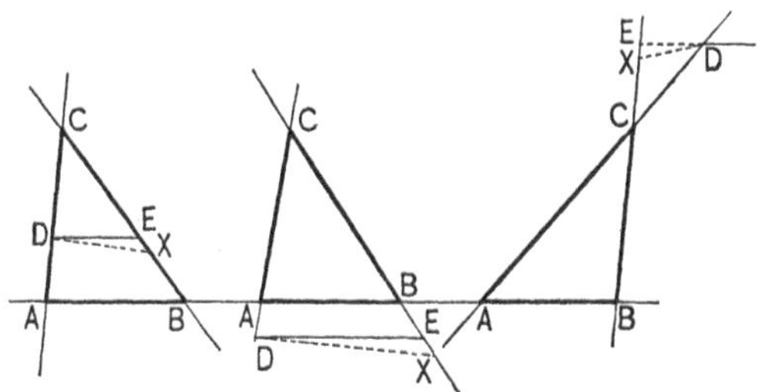

Given   the triangle ABC, and DE so drawn that $AD : DC = BE : EC$.

To prove that $DE \parallel AB$.

Proof. 1. Suppose DE not $\parallel AB$, but that $DX \parallel AB$.

Then $BX : XC = AD : DC$.   Prop. X, cor. 1

2. But this is impossible, for the division of BC in the ratio $AD : DC$ is unique.   Prop. XI, cor.

3. $\therefore DX$ must be identical with DE, and $DE \parallel AB$. The proof is the same for all of the figures.

Exercises. 379. In the above figures, if $AD : DC = BE : EC = m : n$, and if the line through A and E cuts the line through B and D at P, then prove that $AP : PE = BP : PD = m + n : n$.

380. If ex. 379 has been proved, show from it that the centroid of a triangle divides the medians in the ratio of $2 : 1$.

381. Prove prop. XI on the following figures:

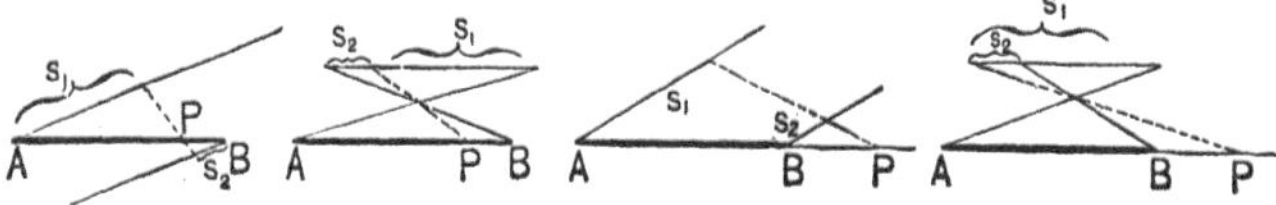

PROPOSITION XIII.

249. Theorem. *If any angle of a triangle is bisected, internally or externally, by a line cutting the opposite side, then the opposite side is divided, internally or externally, respectively, in the ratio of the other sides of the triangle.*

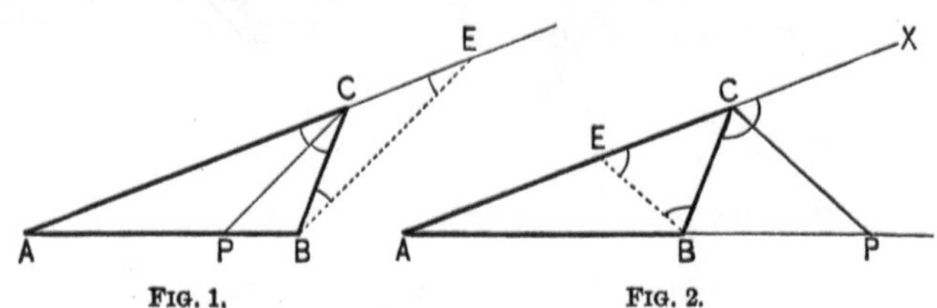

Given $\triangle ABC$, the bisector of $\angle C$ meeting AB at P.

To prove   that $AP : PB = AC : BC$.

Proof.   1.   Let $BE \parallel PC$, meeting AC produced at E, in Fig. 1.
    Then $\angle EBC = \angle PCB = \angle ACP = \angle CEB$.

Given; I, prop. XVII, cor. 2

2. $\therefore BC = CE.$ Why?

3.   But in $\triangle ABE$, $AP : PB = AC : CE$,   Prop. X, cor. 1
    and $\therefore AP : PB = AC : BC.$ Why?
    The proof for Fig. 2 is the same if step 1 is changed
    to $\angle CBE = \angle BCP = \angle PCX = \angle BEC.$

250. Definition.   When a line is divided internally and externally into segments having the same ratio, it is said to be divided **harmonically.**

If the internal and external points of division of AB, in prop. XIII, are P and P', then AB is divided harmonically by P and P'.

Exercise.   382. The hypotenuse of a right-angled triangle is divided harmonically by any pair of lines through the vertex of the right angle, making equal angles with one of its arms.

4. A PENCIL CUT BY ANTIPARALLELS OR BY A CIRCUMFERENCE.

251. **Definitions.** If a pencil of two lines $O - XY$ is cut by two parallel lines AB, MN, and if MN revolves, through a

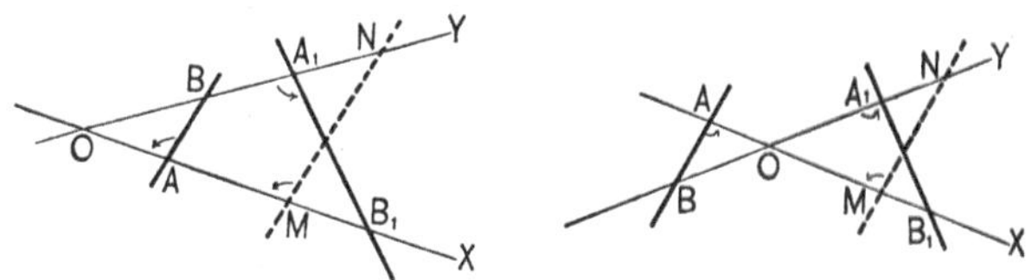

straight angle, about the bisector of $\angle XOY$ as an axis, falling in the position A_1B_1, then AB and A_1B_1 are said to be **antiparallel** to each other.

OA and OA_1 are called *corresponding segments* of the pencil, as are also OB and OB_1. A and A_1 are called *corresponding points*, as are also B and B_1.

Corollary. *If $\angle A = \angle A_1$, in the above figure, then AB and A_1B_1 are antiparallel to each other.*

Exercises. 383. From P, a given point in the side AB of $\triangle ABC$, draw a line to AC produced so that it will be bisected by BC.

384. Investigate ex. 383 when P is on AB produced.

385. If the vertices of $\triangle XYZ$ lie on the sides of $\triangle abc$ so that $x \parallel a$, $y \parallel b$, $z \parallel c$, then X, Y, Z bisect a, b, c.

386. In prop. XIII, suppose $\angle B = \angle A$; also, suppose $\angle B < \angle A$.

387. In any triangle the line joining the feet of the perpendiculars from any two vertices to the opposite sides is antiparallel to the third side.

388. In $\triangle ABC$, suppose that $a \perp c$, and the bisectors of the interior and exterior angles at C meet AB at P_1, P_2. Prove that if a circumference passes through P_1, P_2, and C, (1) P_1P_2 is the diameter, (2) AC is a tangent.

Proposition XIV.

252. Theorem. *If a pencil of two lines is cut by two antiparallel lines, the corresponding segments form a proportion.*

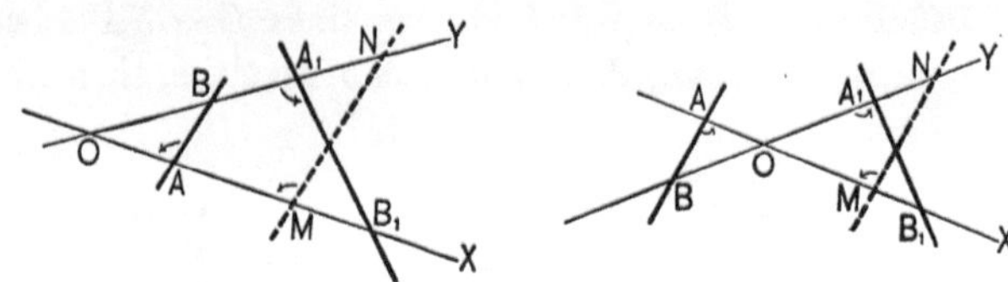

Given the pencil $O - XY$, cut by the antiparallels AB, A_1B_1, A and A_1 being corresponding points.

To prove that $OA : OA_1 = OB : OB_1$.

Proof. 1. Suppose MN the parallel to AB which, revolving, fixed A_1B_1.

Then $OA_1 = OM$, and $OB_1 = ON$. Def. antipar. § 251

2. But $OA : OM = OB : ON$, Prop. X, cor. 2

and $\therefore OA : OA_1 = OB : OB_1$. Substitution

Corollary. *If two antiparallels cut a pencil of two lines, the product of the segments of one line equals the product of the segments of the other.*

Why ? What is meant by "product of two segments" ?

Exercises. 389. In the above figures, $AB : A_1B_1 = OA : OA_1 = OB : OB_1$. (Prop. X, cor. 3, etc.)

390. In the above figures, if A_1 coincides with B, and if $OB = b$, $OA = a$, $OB_1 = b_1$, then $b^2 = ab_1$.

391. If from the vertex of a right-angled triangle a perpendicular p is drawn cutting the hypotenuse c into two segments x, y, adjacent to sides a, b, respectively, then (1) a and p are antiparallels of the pencil b, c; (2) a is a mean proportional between c and x; (3) p is a mean proportional between x and y; (4) $b^2 = cy$, $a^2 = cx$, and $\therefore a^2 + b^2 = c(x + y) = c^2$. (Thus a new proof is found for the Pythagorean proposition.)

Proposition XV.

253. Theorem. *If a pencil of lines cuts a circumference, the product of the two segments from the vertex is constant, whichever line is taken.*

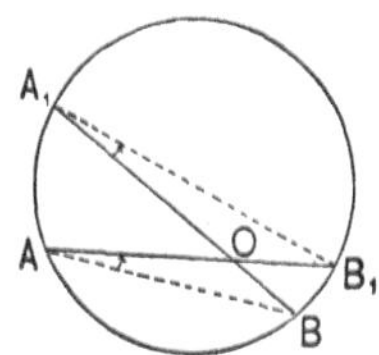

Fig. 1. — The point O on the chord.

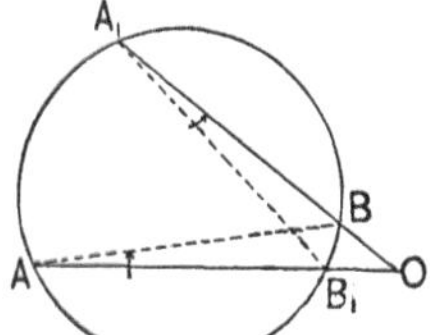

Fig. 2. — The point O on the chord produced.

Given AB_1 and A_1B, two chords, each divided at O into two segments.

To prove that $AO \cdot OB_1 = A_1O \cdot OB.$

Proof. 1.   Suppose AB, A_1B_1 drawn.

Then· $\angle A = \angle A_1.$ Why ?

2. $\therefore AB$ and A_1B_1 are antiparallel,   § 251, cor.

and $\therefore AO \cdot OB_1 = A_1O \cdot OB.$ Prop. XIV, cor.

The proposition is entirely general and should be proved for the following cases.

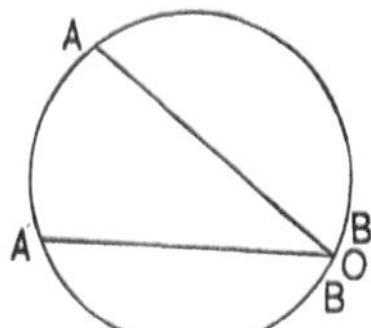

Fig. 3. — The point O at the end of the chord.

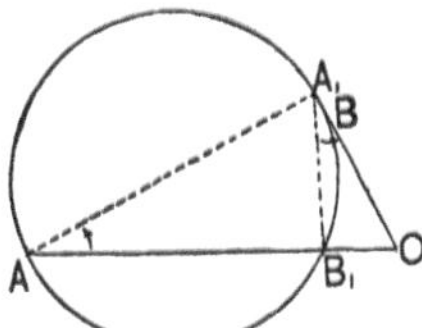

Fig. 4. — Chord A_1B becomes zero.

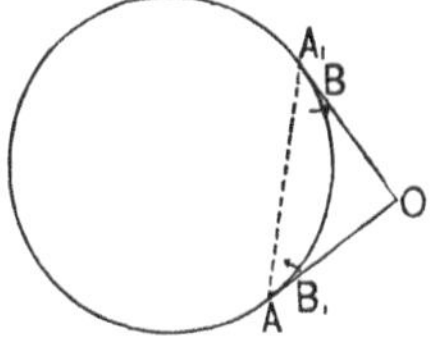

Fig. 5. — Chord AB_1 also becomes zero.

Corollary. *The tangent from the vertex of a pencil to a circumference is a mean proportional between the two segments of any other line of the pencil.*

In Fig. 4, $AO \cdot B_1O = A_1O \cdot BO = BO^2$. Therefore $AO : BO = BO : B_1O.$

Proposition XVI.

254. Theorem. *In the same circle or in equal circles central angles are proportional to the arcs on which they stand.*

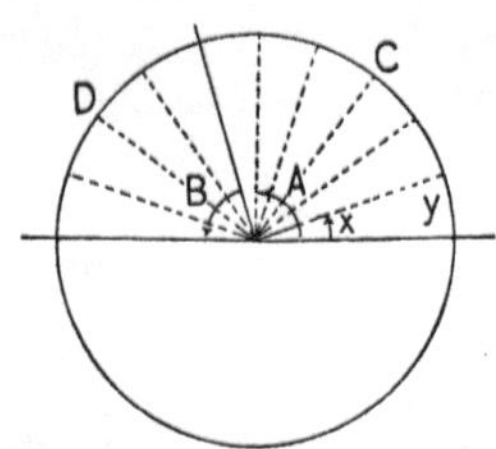

Given A and B, two central angles standing on arcs C and D, respectively.

To prove that $A : B = C : D.$

Proof. 1. If A and B are in different circles, they may be placed in the relative positions shown in the figure.

§ 108, def. ⊙, cor. 2

Suppose A and B divided into equal $\angle$s x,

and suppose $A = nx$, and $B = n'x$.

(In the figure, $n = 6$, $n' = 4$.)

2. Then C is divided into n equal arcs y,

and D "   " n' "   " y.   III, prop. I

3. $\therefore \dfrac{A}{B} = \dfrac{nx}{n'x}$

$$= \frac{n}{n'} = \frac{ny}{n'y} = \frac{C}{D}.$$ Why ?

$$\therefore \frac{A}{B} = \frac{C}{D}.$$ Why ?

Note. The above proof assumes that A and B are commensurable, and hence that they can be divided into equal angles x. The proof on p. 181 is valid if A and B are incommensurable.

255. Proof for incommensurable case.

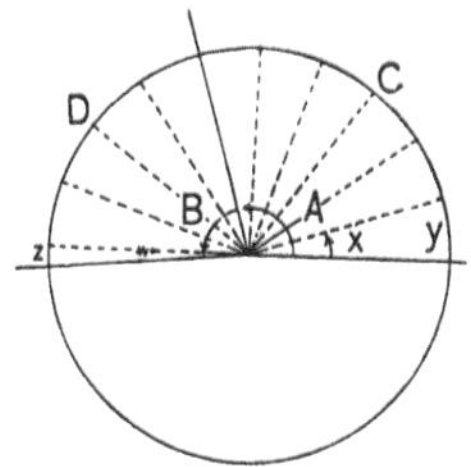

1. Suppose A divided into equal $\measuredangle$ x, and suppose $A = nx$, while $B = n'x + $ some remainder w, such that $w < x$.

   Then C is divided into n equal arcs y, and D is the sum of n' equal arcs $y + $ a remainder z, such that $z < y$.

2. Then B lies between $n'x$ and $(n'+1)\,x$,   Why ?

   and D lies between $n'y$ and $(n'+1)\,y$.   Why ?

3. $\therefore \dfrac{B}{A}$ and $\dfrac{D}{C}$ both lie between $\dfrac{n'}{n}$ and $\dfrac{n'+1}{n}$.   Why ?

   (In the figure, between $\frac{4}{6}$ and $\frac{5}{6}$.)

   And $\therefore \dfrac{B}{A}$ and $\dfrac{D}{C}$ differ by less than $\dfrac{1}{n}$.   Why ?

4. And $\because \dfrac{1}{n}$ can be made smaller than any assumed difference, by increasing n,

   $\therefore$ to assume any difference leads to an absurdity.

5. $\therefore \dfrac{B}{A} = \dfrac{D}{C}$, whence $\dfrac{A}{B} = \dfrac{C}{D}$.

Corollary. *In the same circle or in equal circles sectors are proportional to their angles or to their arcs.*

256. This proposition is often stated,

A central angle is measured by its intercepted arc. See § 180.

5. SIMILAR FIGURES.

257. Definitions. We have (§ 59) roughly defined similar figures as figures having the same shape. But this is unsatisfactory because the word *shape* is not defined. We therefore proceed scientifically to define

    1. *Similar systems of points,* and then

    2. *Similar figures.*

Two systems of points, A_1, B_1, C_1, ... *and* A_2, B_2, C_2, ..., are said to be *similar* when they can be so placed that all lines, A_1A_2, B_1B_2, C_1C_2, ..., joining corresponding points form a pencil whose vertex, O, divides each line into segments having a constant ratio r.

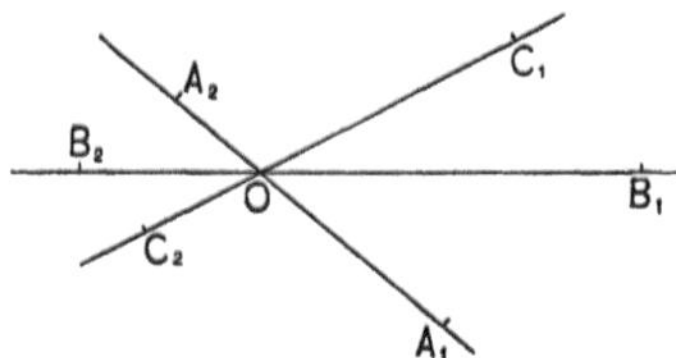

In the figure, $OA_1 : OA_2 = OB_1 : OB_2 = \cdots = r$.

258. *Two figures* are said to be similar when their systems of points are *similar*.

The symbol of similarity ∽, already mentioned, is due to Leibnitz. It is derived from the letter S.

The following are illustrations of similar figures involving circles:

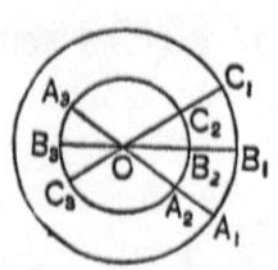

Concentric circles.

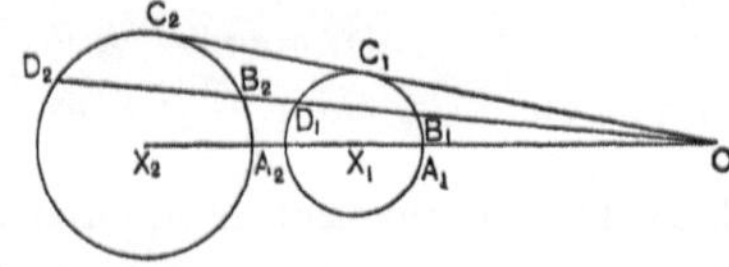

Any circles.

The following are illustrations of similar rectilinear figures :

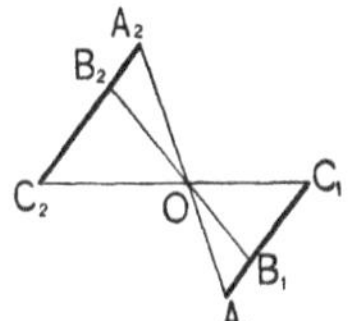 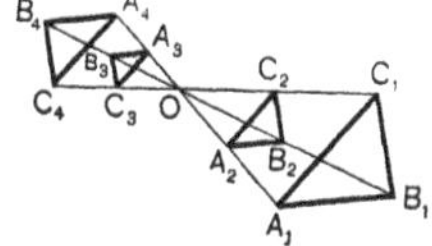 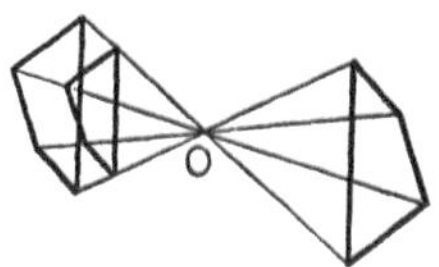

Any line-segments.   Four similar triangles.   Three similar quadrilaterals.

259. When two similar figures are so placed that lines through their corresponding points form a pencil, they are said to be placed *in perspective*, and the vertex of that pencil is called their *center of similitude.*

The figures above and on p. 182 are placed in perspective, and in each case O is the center of similitude.

Two similar figures may evidently be so placed that the center of similitude will fall within both, or between them, or on the same side of both, as is seen in the above illustrations.

260. *Two systems of points, A_1, B_1, C_1, ... and A_2, B_2, C_2, ..., are said to be* **symmetric** *with respect to a center O when all lines, A_1A_2, B_1B_2, C_1C_2, ..., are bisected by O.*

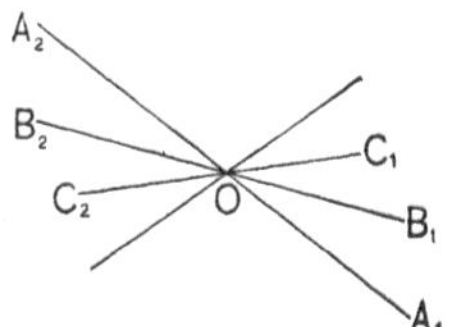

261. Two *figures* are said to be **symmetric** *with respect to a center* when their systems of points are symmetric with respect to that center.

E.g. in the figure, $\triangle A_1B_1C_1$, $A_2B_2C_2$ are symmetric with respect to O.

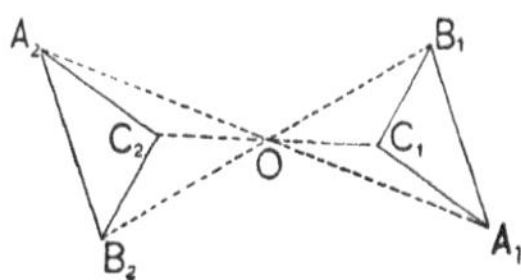

262. In similar figures, if the ratio, r, known as the **ratio of similitude,** is 1, the figures are evidently symmetric with respect to a center.   Hence Central Symmetry is a special case of Similar Figures in Perspective.

The term Center of Similitude is due to Euler.

Corollaries.   1. *Congruent figures are similar.*

For if made to coincide, any point in their plane is evidently a center of similitude, the ratio of similitude being 1.   Or, they may be placed in a position of central symmetry.

2. *To any point in a system there is one and only one corresponding point of a similar system with respect to a given center.*

If A_1 and A_2 are corresponding points in two similar systems in perspective, and O is the center of similitude, then every point P_1 on OA_1 has a unique corresponding point P_2 on OA_2.

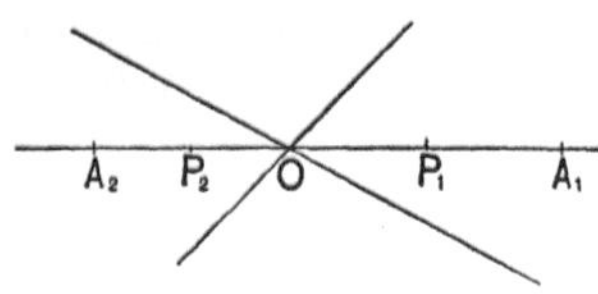

For $OA_1 : OA_2 = OP_1 : OP_2$,

$\therefore OP_2$ is unique.   § 237, cor. 1

Exercises.   **392.** What is the limit of $1/x$ as x increases indefinitely ? of $1/(1 + x)$ as $x \doteq 0$?   as $x \doteq 1$?

393. In $\triangle ABC$, P is any point in AB, and Q is such a point in CA that $CQ = PB$; if PQ and BC, produced if necessary, meet at X, prove that $CA : AB = PX : QX$.   (From P draw a line $\parallel AC$.)

394. In the annexed figure of a "Diagonal Scale," AB is 1 centimeter.   Show how, by means of the scale and a pair of dividers, to lay off 1 millimeter, 0.5 millimeter, 0.3 millimeter, etc.   On what proposition or corollary does this measurement of fractions of a millimeter depend ?

395. $ABCD$ is a parallelogram; from A a line is drawn cutting BD in E, BC in F, and DC produced in G.   Prove that AE is a mean proportional between EF and EG.

396. ABC is a triangle, and through D, any point in c, DE is drawn $\parallel a$ to meet b in E; through C, CF is drawn $\parallel EB$ to meet c produced in F.   Prove that AB is a mean proportional between AD and AF.

Proposition XVII.

263. Theorem. *Two triangles are similar if they have two angles of the one equal to two angles of the other, respectively.*

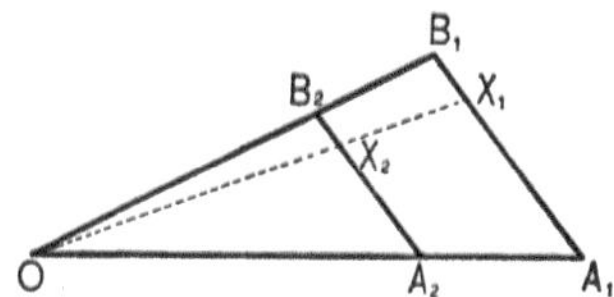

Given   the $\triangle A_1B_1C_1$, $A_2B_2C_2$, with $\angle A_1 = \angle A_2$, $\angle C_1 = \angle C_2$.

To prove that $\triangle A_1B_1C_1 \backsim \triangle A_2B_2C_2$.

Proof. 1. Place one $\triangle$ on the other so that $\angle C_1$ coincides with $\angle C_2$, as at O, OA_2 falling on OA_1. Let OX_2X_1 be *any* line through O, cutting A_2B_2 at X_2, and A_1B_1 at X_1.

Then $\because \angle A_2 = \angle A_1$,

$\therefore A_2B_2 \parallel A_1B_1$.   I, prop. XVI, cor. 1

2. $\therefore OA_1 : OA_2 = OB_1 : OB_2 = OX_1 : OX_2 = \cdots\cdots = r$.
Prop. X, cor. 2

3. And *all* points on OA_1 and OB_1 have their corresponding points on OA_2 and OB_2, respectively.
§ 262, cor. 2

4. $\therefore$ the $\triangle$ are similar, O being the center of similitude.
§ 258

Corollaries. 1. *Mutually equiangular triangles are similar.*

2. *If two triangles have the sides of the one respectively parallel or perpendicular to the sides of the other, they are similar.*

For by § 86, cor. 5, they can be proved to be mutually equiangular.

Proposition XVIII.

264. Theorem. *If two triangles have one angle of the one equal to one angle of the other, and the including sides proportional, the triangles are similar.*

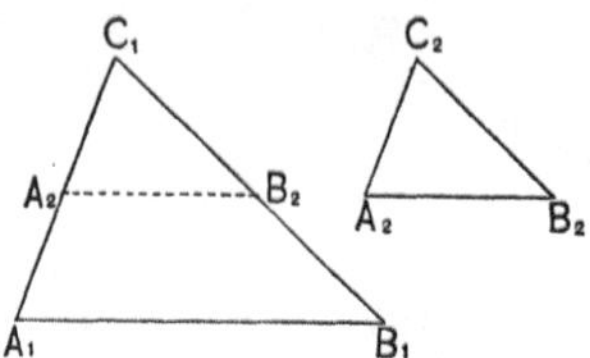

Given $\triangle A_1B_1C_1$, $A_2B_2C_2$, such that $\angle C_1 = \angle C_2$ and $a_1 : a_2 = b_1 : b_2$.

To prove that $\triangle A_1B_1C_1 \backsim \triangle A_2B_2C_2$.

Proof. 1. $\because \angle C_2 = \angle C_1$, $\triangle A_2B_2C_2$ can be placed on $\triangle A_1B_1C_1$ so that C_2 falls at C_1, B_2 on a_1, and A_2 on b_1.

   2. Then $\because C_1A_2 : C_1A_1 = C_1B_2 : C_1B_1$,

$$\therefore A_2B_2 \parallel A_1B_1. \qquad \text{Props. XII, V}$$

   3. $\therefore \triangle A_1B_1C_1$ and $\triangle A_2B_2C_1$ are mutually equiangular.

I, prop. XVII, cor. 2

   4. $\therefore \triangle A_1B_1C_1 \backsim \triangle A_2B_2C_1$ and its congruent $\triangle A_2B_2C_2$.

Prop. XVII, cor. 1

Exercises. 397. ABC, DBA are two triangles with a common side AB. If P is any point on AB, and $PX \parallel AC$, and $PY \parallel AD$, meeting BC and BD in X and Y, respectively, prove that $\triangle YBX \backsim \triangle DBC$.

398. $ABCD$ is a quadrilateral. Prove that if the bisectors of $\angle A$, C meet on diagonal BD, then the bisectors of $\angle B$, D will meet on diagonal AC.

399. Construct a triangle, having given the base, the vertical angle, and the ratio of the remaining sides. (*Intersection of loci* and prop. XIII.)

400. In $\triangle ABC$, CM is a median; $\angle BMC$, CMA are bisected by lines meeting a and b in R and Q, respectively. Prove that $QR \parallel AB$.

Proposition XIX.

265. Theorem. *If two triangles have their sides proportional, they are similar.*

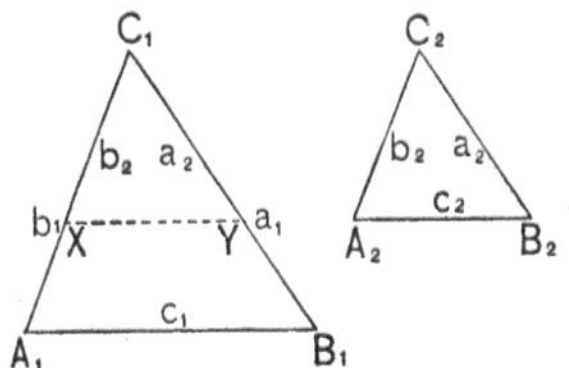

Given $\triangle$ $A_1B_1C_1$, $A_2B_2C_2$ such that $a_1 : a_2 = b_1 : b_2 = c_1 : c_2$.

To prove that $\triangle A_1B_1C_1 \backsim \triangle A_2B_2C_2$.

Proof.   1. On C_1A_1, C_1B_1 lay off $C_1X = b_2$, and $C_1Y = a_2$, and draw XY.

Then $\because a_1 : a_2 = b_1 : b_2$, and $\angle C_1 \equiv \angle C_1$,

2. $\therefore \triangle XYC_1 \backsim \triangle A_1B_1C_1$.   Prop. XVIII

3. But $a_1 : a_2 = c_1 : c_2$,   Given

   and $a_1 : a_2 = c_1 : XY$.   Prop. X, cor. 3

4. $\therefore c_1 : c_2 = c_1 : XY$,   Why ?

   and $c_2 = XY$.   § 237, cor. 2

5. $\therefore \triangle XYC_1 \cong \triangle A_2B_2C_2$,   I, prop. XII

   and $\therefore \triangle A_1B_1C_1 \backsim \triangle A_2B_2C_2$.   Steps 2, 5

Exercises. 401. The product of the two segments of any chord drawn through a given point within a circle equals the square of half the shortest chord that can be drawn through that point.

402. If P is a point on AB produced, the tangents from P to all circumferences through A and B are equal, and hence such points are concyclic.

403. If from any point P on the side CA of a right-angled triangle ABC, PQ is drawn perpendicular to the hypotenuse AB at Q, then $AP \cdot AC = AQ \cdot AB$. Suppose P to be taken (1) at C ; (2) at A ; (3) on AC produced.

PROPOSITION XX.

266. Theorem. *Similar triangles have their corresponding sides proportional and their corresponding angles equal.*

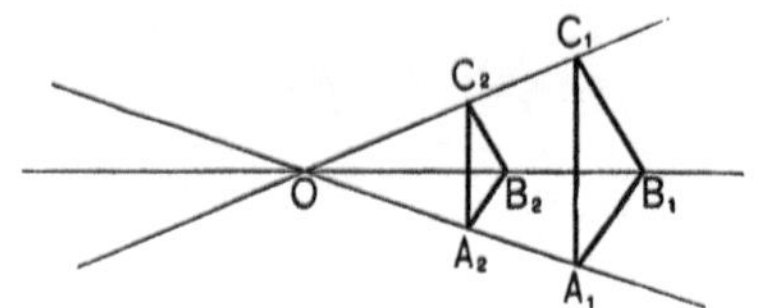

Given two similar triangles, $A_1B_1C_1$, $A_2B_2C_2$, A_1 corresponding to A_2, B_1 to B_2, C_1 to C_2.

To prove that $A_1B_1 : A_2B_2 = B_1C_1 : B_2C_2 = \ldots$ and that $\angle B_1 = \angle B_2 \ldots$

Proof. 1. Suppose the $\triangle$ placed in perspective.

Then $OA_1 : OA_2 = OB_1 : OB_2 = OC_1 : OC_2$. § 258

2. $\therefore A_1B_1 \parallel A_2B_2$, and so for other sides.

Props. XII, V

3. $\therefore \angle OB_1A_1 = \angle OB_2A_2$, and $\angle C_1B_1O = \angle C_2B_2O$.

I, prop. XVII, cor. 2

4. $\therefore \angle B_1 = \angle B_2$, and so for other angles. Ax. 2

5. Also, $OB_1 : OB_2 = A_1B_1 : A_2B_2$,

$= B_1C_1 : B_2C_2$. Prop. X, cor. 3

6. $\therefore A_1B_1 : A_2B_2 = B_1C_1 : B_2C_2$, and so for other sides.

NOTE. This is the converse of props. XVII, XIX.

COROLLARIES. 1. *The corresponding altitudes of two similar triangles have the same ratio as any two corresponding sides.*
Why?

2. *The corresponding sides of similar triangles are opposite the equal angles.*

In what step is this proved?

267. Summary of Propositions concerning Similar Triangles.

Two triangles are similar if

1. (a) *Two angles of the one equal two angles of the other.*
Prop. XVII

   (b) *They are mutually equiangular.*   Prop. XVII, cor. 1

   (c) *The sides of the one are parallel to the sides of the other.*   Prop. XVII, cor. 2

   (d) *The sides of the one are perpendicular to the sides of the other.*   Prop. XVII, cor. 2

2. *One angle of the one equals one angle of the other and the including sides are proportional.*   Prop. XVIII

3. *Their corresponding sides are proportional.*   Prop. XIX

If two triangles are similar,

1. *They are mutually equiangular.*   Prop. XX

2. *Their corresponding sides are proportional.*   Prop. XX

3. *Their corresponding altitudes are proportional to their corresponding sides.*   Prop. XX, cor. 1

268. It should further be observed that, in general,

Three conditions determine congruence.   (See § 90.)

Two conditions determine similarity.

For these conditions are

1. Two angles equal.   (Prop. XVII.)

2. One angle and one ratio.   (Prop. XVIII.)

3. Two ratios ; for if the sides are a, b, c, and a', b', c', then if $\dfrac{a}{b} = \dfrac{a'}{b'}$, and $\dfrac{a}{c} = \dfrac{a'}{c'}$, the $\triangle$ are similar, since $\dfrac{b}{c}$ must also equal $\dfrac{b'}{c'}$.

Exercises. 404. If X is any point in the side a, or a produced, of $\triangle ABC$, and if r_b and r_c are the radii of circles circumscribed about $\triangle ABX$ and $\triangle AXC$, respectively, then $r_b : r_c = c : b$. (Join the centers and prove two triangles similar.)

405. If one of the parallel sides of a trapezoid is double the other, prove that the diagonals intersect one another in a point of trisection.

Proposition XXI.

269. Theorem. *If two polygons are mutually equiangular and have their corresponding sides proportional, they are similar.*

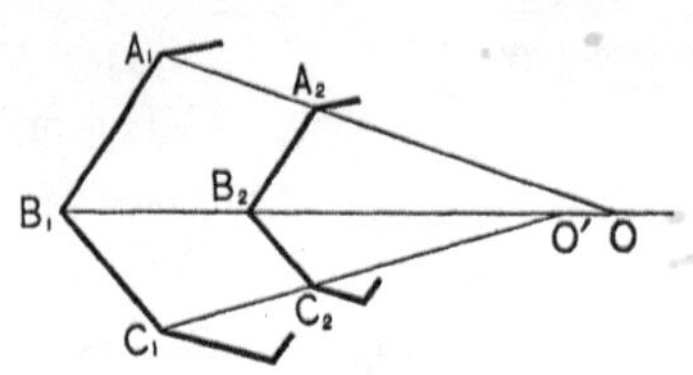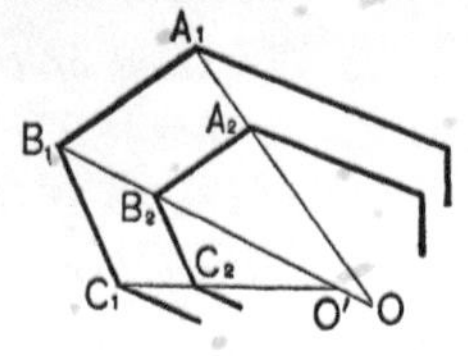

Given two polygons, $A_1B_1C_1 \ldots$ and $A_2B_2C_2 \ldots$, such that
$$\angle A_1 = \angle A_2, \ \angle B_1 = \angle B_2, \ldots,$$
and $\quad A_1B_1 : A_2B_2 = B_1C_1 : B_2C_2 = \ldots$

To prove that $\quad A_1B_1C_1 \ldots \backsim A_2B_2C_2 \ldots$

Proof. 1. Place $A_2B_2 \parallel A_1B_1$. Then $\because$ the $\angle\!\!\!\!s$ of one polygon $=$ the corresponding $\angle\!\!\!\!s$ of the other, the remaining sides may be made parallel respectively.

I, prop. XVII, cor. 5

2. If $A_1B_1 > A_2B_2$, then $B_1C_1 > B_2C_2, \ldots$, because the ratios are equal.

3. Draw $A_1A_2, B_1B_2, \ldots$ Then $A_1B_1B_2A_2$ is not a $\square$; also $B_1C_1C_2B_2$, etc.; and A_1A_2 meets B_1B_2 as at O, B_1B_2 meets C_1C_2 as at O', etc. I, prop. XXIV

4. But $B_1O' : B_2O' = B_1C_1 : B_2C_2$
$$= A_1B_1 : A_2B_2 = B_1O : B_2O,$$

Prop. X, cor. 3

which is impossible unless O and O' coincide.

Prop. XI, cor.

5. $\therefore$ the two figures are similar, and O is the center of similitude. § 258

In step 2, if $A_1B_1 = A_2B_2$, then $B_1C_1 = B_2C_2$, ..., and the polygons are congruent and therefore similar.   § 262, cor. 1

COROLLARIES. 1. *If two polygons are similar, they are mutually equiangular and their corresponding sides are proportional.*

For if placed in perspective as on p. 190,

1. $OA_1 : OA_2 = OB_1 : OB_2$.   § 258
2. ∴ $A_1B_1 \parallel A_2B_2$, and so for other sides.   Prop. XII
3. ∴ $\angle B_1 = \angle B_2$, and so for other angles.   I, prop. XVII, cor. 5
4. Also $A_1B_1 : A_2B_2 = B_1O : B_2O = B_1C_1 : B_2C_2 = $ .....   Prop. X, cor. 3

2. *Polygons similar to the same polygon are similar to each other.*

For they have angles equal to those of the third polygon, and the ratios of their sides equal the ratios of the sides of the third polygon.

3. *The perimeters of similar polygons have the same ratio as the corresponding sides.*

For by cor. 1, $A_1B_1 : A_2B_2 = B_1C_1 : B_2C_2 = $ ..... $= r$. ∴ $A_1B_1 + B_1C_1 + $ ..... $: A_2B_2 + B_2C_2 + $ ..... $= r$. (Why ?)

4. *Two similar polygons can be divided into the same number of triangles similar each to each, and similarly placed.*

For O and O' coincide, and the figures can be placed having O within each. The triangles A_1OB_1, A_2OB_2 are then similar, by prop. XVII.

Exercises. 406. If from a point outside a circle a pair of tangents and a secant are drawn, the quadrilateral formed by joining in succession the four points thus determined on the circumference has the rectangles of its opposite sides equal.

407. AB is a diameter, and from A a line is drawn to cut the circumference in C and the tangent from B in D. Prove that the diameter is the mean proportional between AC and AD.

408. In $\square ABCD$, P, Q are points in a line parallel to AB; PA and QB meet at R, and PD and QC meet at S. Prove that $RS \parallel AD$.

409. Chords AB, CD are produced to meet at P, and PF is drawn parallel to DA to meet CB produced in F. Prove that PF is the mean proportional between FB and FC.

Proposition XXll.

270. Theorem. *In a right-angled triangle the perpendicular from the vertex of the right angle to the hypotenuse divides the triangle into two triangles which are similar to the whole and to each other.*

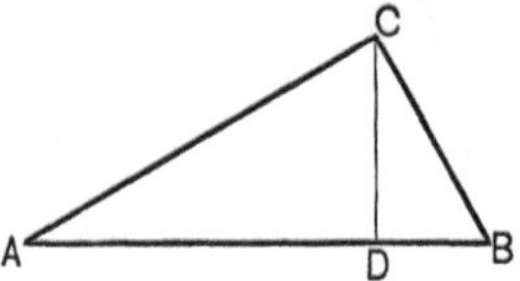

Given $\triangle ABC$, with $\angle C$ a right angle, and $CD \perp AB$.

To prove that (1) $\triangle ACD \backsim \triangle ABC$.

(2) $\triangle CBD \backsim \triangle ABC$.

(3) $\triangle ACD \backsim \triangle CBD$.

Proof. 1. $\because \angle CDA = \angle ACB,$ Why?

and $\angle A \equiv \angle A,$

$\therefore \triangle ACD \backsim \triangle ABC$, which proves (1).

Prop. XVII

2. Similarly $\triangle CBD \backsim \triangle ABC$, which proves (2).

Prop. XVII

3. $\therefore \triangle ACD \backsim \triangle CBD$, which proves (3).

Prop. XXI, cor. 2

Corollaries. 1. *Either side of a right-angled triangle is the mean proportional between the hypotenuse and its segment adjacent to that side.*

For from step 1, $AB : AC = AC : AD$; and from 2, $AB : BC = BC : DB$.

2. *The perpendicular from the vertex of the right angle to the hypotenuse is the mean proportional between the segments of the hypotenuse.*

For from step 3, $AD : CD = CD : DB$.

EXERCISES.

410. Prove the converse of prop. XXII : If the perpendicular drawn from the vertex of a triangle to the base is the mean proportional between the segments of the base, the triangle is right-angled.

411. Prove that any chord of a circle is the mean proportional between its projection on the diameter from one of its extremities, and the diameter itself.

412. In the figure on p. 192, if AD represents three units, and DB represents one unit, what number is represented by CD ?

413. Prove that if a perpendicular is let fall from any point on a circumference, to any diameter, it is the mean proportional between the segments into which it divides that diameter.

414. Prove that if two fixed parallel tangents are cut by a variable tangent, the rectangle of the segments of the latter is constant.

415. Through any point in the common chord of two intersecting circumferences two chords are drawn, one in each circle. Prove that the four extremities of these chords are concyclic.

416. If the bisectors of the interior and exterior angles at B, in the figure of prop. XXII, meet b at F and E, respectively, prove that BC is the mean proportional between FC and CE.

417. Calculate each of the segments into which the bisectors of the angles of a triangle divide the opposite sides, the lengths of the sides being 9 in., 12 in., and 15 in., respectively.

418. From the points A, B, on a line AB, 25 in. long, perpendiculars AC, BD are erected such that $AC = 13$ in., $BD = 7$ in. On AB the point O is taken such that $\angle BOD = \angle COA$. Calculate the distances AO, OB.

419. Given a trapezoid $ABCD$, with the non-parallel sides AD, BC divided at E, F, respectively, in the ratio of 2 to 3, to calculate the length of EF, knowing that $AB = 12.45$ in., and $DC = 38.5$ in.

420. Calculate the sides of a right triangle, knowing that their respective projections on the hypotenuse are 2.88 in. and 5.12 in.

421. The two sides of a right triangle are respectively 10 in. and 24 in. Required the lengths of their projections on the hypotenuse, and the distance of the vertex of the right angle from the hypotenuse. (To 0.001.)

422. The two sides of a right triangle are respectively 3.128 in. and 4.275 in. Required the lengths of the two segments into which the bisector of the right angle divides the hypotenuse. (To 0.001.)

6. PROBLEMS.

Proposition XXIII.

271. Problem. *To divide a line-segment into parts proportional to the segments of a given line.*

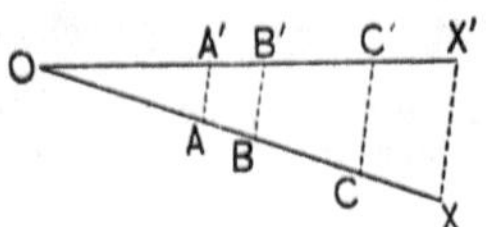

Given   the line OX', and the line OX divided into segments OA, AB, ...

Required   to divide OX' into segments proportional to OA, AB, ...

Construction. 1. Placing the lines oblique to each other at a common end-point O, draw XX'.   § 28

     2. From A, B, ... draw lines ‖ XX', cutting OX' at A', B', ...   I, prop. XXXIII

     Then OX' is divided as required.

Proof.   ∵ OX, OX' are two transversals of a pencil of ‖'s, the corresponding segments are in proportion.   Prop. X

Corollaries. 1. *A given line can be divided into parts proportional to any number of given lines.*

For that number of given lines may be laid off as OA, AB, BC, ... on OX.

2. *A line can be divided into any number of equal parts.*

Note. While a straight line can be divided into any number of equal parts, by means of the straight edge and the compasses, a circumference cannot be divided into 7, 9, 11, 13, and, in general, any prime number of equal parts beyond 5. The exceptions are noted in Book V.

Proposition XXIV.

272. **Problem.** *To find the fourth proportional to three given lines.*

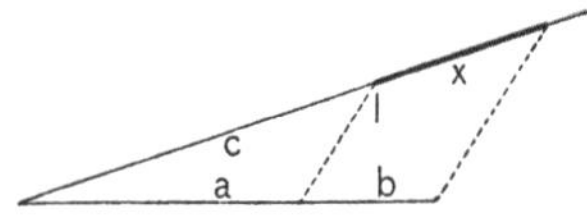

Given   three lines, a, b, c.

Required   to find x such that $a : b = c : x$.

Construction. 1. From the vertex of a pencil of two lines, with the compasses lay off a, b, in order, on one line, and c on the other line.

2. Join the end-points of a, c, remote from the vertex, by l.   § 28

3. From the end-point of b, remote from a, draw a line parallel to l.   I, prop. XXXIII

This will cut off x, the line required.

Proof. $a : b = c : x.$ Prop. X, cor. 1

273. **Definition.** If $a : b = b : x$, x is called the *third proportional* to a and b.

Corollary. *The third proportional to two given lines can be found.*

For to find x such that $a : b = b : x$, make $c = b$ in the above solution.

Exercises. **423.** The problem admits of a considerable variation of the figure, as suggested by the figure given in ex. 383. Invent another solution from this suggestion.

424. How many inches in the fourth proportional to lines respectively 2 in., 3 in., 5 in. long? In the third proportional to lines respectively 2 in., 7 in. long?

Proposition XXV.

274. Problem. *To find the mean proportional between two given lines.*

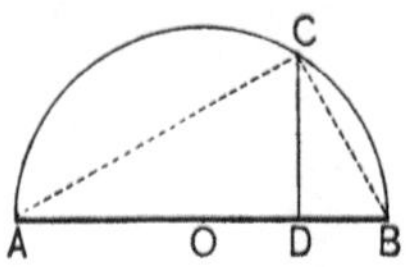

Given   two lines, AD, DB.

Required to find the mean proportional between them.

Construction. 1. Placing AD, DB end to end in the same line, bisect AB at O.   I, prop. XXXI

    2. With center O and radius OB, describe a circle.

§ 109

    3. From D draw $DC \perp AB$, to meet circumference at C.   I, prop. XXIX

    Then CD is the mean proportional.

Proof. $AD : CD = CD : DB$. Prop. XXII, cor. 2, and § 238

275. Definition. A line is said to be divided in **extreme and mean ratio** by a point when one of the segments is the mean proportional between the whole line and the other segment.

Thus, AB is divided internally in extreme and mean ratio at P, if $AB : AP = AP : PB$; and externally in such ratio at P', if $AB : AP' = AP' : P'B$.

To say that $AB : AP = AP : PB$ is merely to say that $AP^2 = AB \cdot PB$. This division is often known as the **Golden Section** or the **Median Section**.

If the student understands quadratic equations he will see that if the length of AB is 6, and if $AP = x$, then $PB = 6 - x$, and

$$\because AP^2 = AB \cdot PB,$$

$$\therefore x^2 = 6(6 - x),$$

or $x^2 + 6x - 36 = 0$. Solving, $x = -3 \pm 3\sqrt{5}$.

Proposition XXVI.

276. Problem. *To divide a line in extreme and mean ratio.*

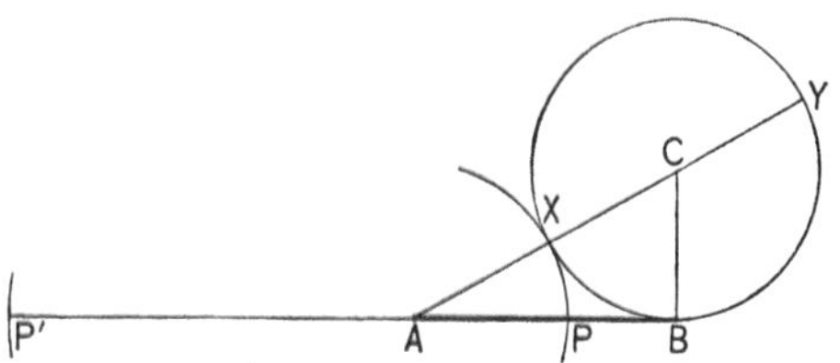

Given the line AB.

Required to divide AB in extreme and mean ratio; *i.e.* to find P such that $AB \cdot PB = AP^2$.

Construction. 1. Draw $CB \perp AB$ and $= \tfrac{1}{2} AB$.

2. Describe a $\odot$ with center C and radius CB.

3. Draw AC cutting the circumference in X and Y.

4. Describe two arcs with center A and radii AX and AY, thus fixing points P, P'.

These are the required points.

Proof for point P.

$$AB^2 = AX \cdot AY$$
$$= AP\,(AX + XY)$$
$$= AP\,(AP + AB)$$
$$= AP^2 + AP \cdot AB.$$
$$\therefore AB\,(AB - AP) = AP^2.$$
$$\therefore AB \cdot PB = AP^2.$$

Proof for point P'.

$$AB^2 = AY \cdot AX$$
$$= P'A\,(AY - XY)$$
$$= P'A\,(P'A - AB)$$
$$= P'A^2 - AB \cdot P'A.$$
$$\therefore AB\,(AB + P'A) = P'A^2.$$
$$\therefore AB \cdot P'B = P'A^2.$$

$\therefore AB$ is divided internally at P and externally at P' in Golden Section.

It should be noticed that if the sense of the lines as positive or negative is considered (that is, considering $AP = -PA$), the above solutions would be identical if X and Y were interchanged, and P' substituted for P.

Proposition XXVII.

277. Problem. *On a given line-segment as a side corresponding to a given side of a given polygon, to construct a polygon similar to that polygon.*

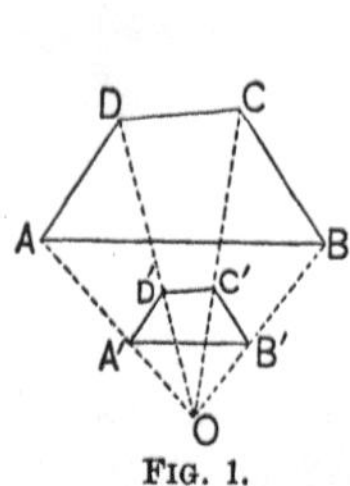

Fig. 1.

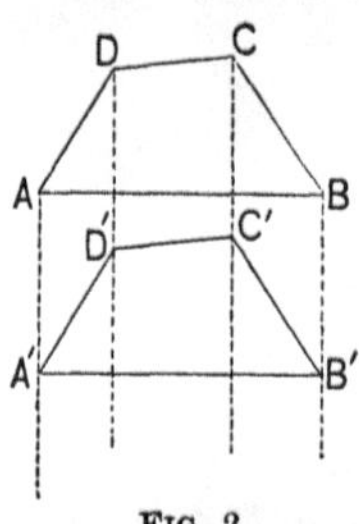

Fig. 2.

Given the polygon $ABCD$ and the line-segment $A'B'$.

Required to construct on $A'B'$ as a side corresponding to AB, a polygon $A'B'C'D' \backsim ABCD$.

Construction. 1. In Fig. 1, place $A'B' \parallel AB$.　I, prop. XXXIII

2. Draw AA', BB', meeting at O; draw OC, OD.　§ 28

3. Draw $B'C' \parallel BC$, $C'D' \parallel CD$.　　　I, prop. XXXIII

4. Draw $D'A'$.　Then $A'B'C'D' \backsim ABCD$.

Proof. 1. $\because OA : OA' = OB : OB' = OC : OC' = OD : OD'$, § 244

$$\therefore D'A' \parallel DA. \qquad \text{Prop. XII}$$

2. $\because A'B' : AB = OB' : OB = B'C' : BC = \cdots\cdots,$

Prop. X, cor. 3

and $\angle C'B'A' = \angle CBA$, and so for the other $\angle\!s$,

I, prop. XVII, cor. 2; ax. 3

$$\therefore A'B'C'D' \backsim ABCD. \qquad \text{Prop. XXI}$$

If $A'B' = AB$, as in Fig. 2, draw from C, D $\parallel$'s to AA'; otherwise the construction is as above.　It is left to the student to prove $D'A' \parallel DA$, and $A'B'C'D' \cong ABCD$.　$\therefore A'B'C'D' \backsim ABCD$ by § 262, cor. 1.

BOOK V.—MENSURATION OF PLANE FIGURES.
REGULAR POLYGONS AND THE CIRCLE.

1. MENSURATION OF PLANE FIGURES.

PROPOSITION I.

278. Theorem. *Two rectangles having equal altitudes are proportional to their bases.*

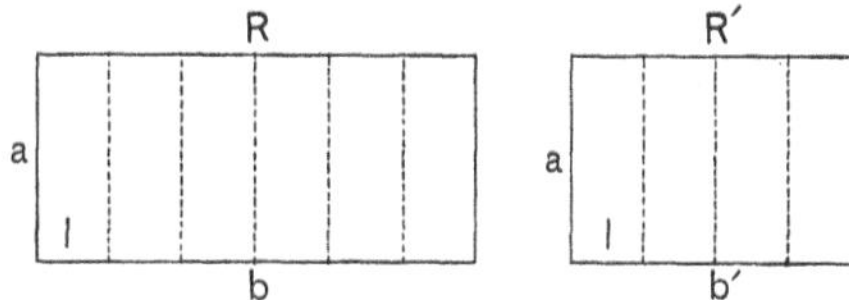

Given two rectangles R and R', with altitude a, and with bases b, b', respectively.

To prove that $R : R' = b : b'$.

Proof. 1. Suppose b and b' divided into equal segments, l, and suppose $b = nl$, and $b' = n'l$.

(In the figures, $n = 6$, $n' = 4$.)

Then if ⊥s are erected from the points of division,

$$R = n \text{ congruent rectangles } al,$$

and $R' = n'$ " " "

2. $\quad \therefore \dfrac{R}{R'} \equiv \dfrac{n \cdot al}{n' \cdot al} = \dfrac{n}{n'} = \dfrac{b}{b'}.$ Why ?

NOTE. The above proof assumes that b and b' are commensurable, and hence that they can be divided into equal segments l. The proposition is, however, entirely general. The proof on p. 200 is valid if b and b' are incommensurable.

199

279. Proof for incommensurable case.

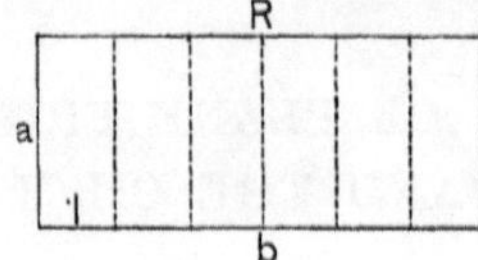
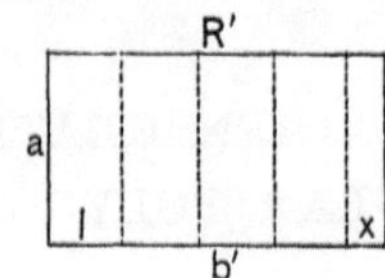

1. Suppose b divided into equal segments l,

   and suppose $b = nl$,

   while $b' = n'l +$ some remainder x,

   such that $x < l$.

   Then if $\perp$'s are erected from the points of division,

        $R = n$ congruent rectangles al,

   and $R' = n'$ "   " $al +$ a remainder ax,

   such that $ax < al$.

2. Then b' lies between $n'l$ and $(n' + 1) l$,   Why?

       (In the figure, between $4\,l$ and $5\,l$.)

   and R' lies between $n' \cdot al$ and $(n' + 1) \cdot al$.   Why?

3. $\therefore \dfrac{b'}{b}$ and $\dfrac{R'}{R}$ both lie between $\dfrac{n'}{n}$ and $\dfrac{n' + 1}{n}$,   Why?

       (In the figure, between $\frac{4}{8}$ and $\frac{5}{8}$.)

   and $\therefore$ they differ by less than $\dfrac{1}{n}\cdot$ Why?

       (In the figure, by less than $\frac{1}{8}$.)

4. And $\therefore \dfrac{1}{n}$ can be made smaller than any assumed

   difference, by increasing n,

   $\therefore$ to assume any difference leads to an absurdity.

5. $\therefore \dfrac{b'}{b} = \dfrac{R'}{R}$, whence $\dfrac{R}{R'} = \dfrac{b}{b'}\cdot$

NOTE. The proof will be noticed to be essentially that of pp. 171, 181.

Corollaries. 1. *Rectangles having equal bases are proportional to their altitudes.*

For they can be turned through 90° so as to interchange base and altitude.

2. *Triangles having equal altitudes are proportional to their bases; having equal bases, to their altitudes.* (Why ?)

3. *Parallelograms having equal bases are proportional to their altitudes; having equal altitudes, to their bases.*

Proposition II.

280. Theorem. *Two rectangles have the same ratio as the products of (the numerical measures of) their bases and altitudes.*

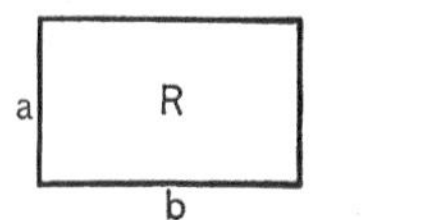

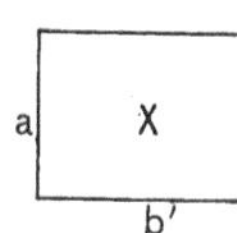

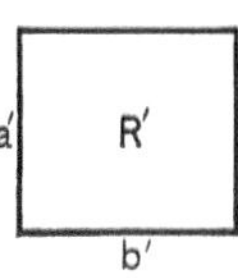

Given two rectangles R, R', with bases b, b', and altitudes a, a', respectively.

To prove that $R : R' = ab : a'b'$.

Proof. 1. Let X be a rectangle of altitude a and base b'.

Then $$\frac{R}{X} = \frac{b}{b'},$$ Prop. I

and $$\frac{X}{R'} = \frac{a}{a'}.$$ Prop. I, cor. 1

2. $$\therefore \frac{R}{R'} = \frac{ab}{a'b'},$$ by multiplying corresponding members of the two equations. Ax. 6

Note. Thus again appears the relation between geometry and algebra set forth in § 221, that to the product of two numbers corresponds the rectangle of two lines.

281. Definition. To **measure** a surface is to find its ratio to some unit. The unit of measure, multiplied by this ratio, is called the *area*.

Thus, in a surface 4 ft. long by 2 ft. broad, the ratio of the surface to 1 sq. ft. is 8, and 8 sq. ft. is the area.

Corollaries. 1. *Parallelograms (or triangles) have the same ratio as the products of their bases and altitudes.* (Why?)

For a parallelogram equals a rectangle of the same base and the same altitude, II, prop. I, cor. 1. See also (for the triangle) II, prop. II, cor. 1.

2. *The area of a rectangle equals the product of its base and altitude.*

That is, the number which represents its square units of area is the product of the two numbers which represent its base and altitude.

For in prop. II, if $R' = 1$, the square unit of area, then a' and b' must each equal 1, the unit of length. Hence $R/1 = ab/1$, or $R = ab$.

3. *The area of a parallelogram equals the product of its base and altitude ; of a triangle, half that product.*

See the proof under cor. 1.

4. *The area of a square equals the second power of its side.*

This is the **reason** that the *second power* of a *number* is called its *square*.

5. *The area of a trapezoid equals the product of its altitude and half the sum of its bases.* (Why?)

See II, prop. III.

Exercises. 425. Prove that any quadrilateral is divided by its interior diagonals into four triangles which form a proportion.

426. ABC is a triangle, and P is any point in BC; from P are drawn two parallels to CA, BA, meeting AB, AC in X, Y, respectively. Prove that $\triangle AXY$ is a mean proportional between $\triangle BPX$ and $\triangle PCY$. Investigate when P is on CB produced.

427. Suppose D, E, the mid-points of sides b, a of $\triangle ABC$, to be joined ; draw AE and BD, intersecting at O. Prove that $\triangle BEO$ is a mean proportional between $\triangle DOE$ and ABO. Investigate when $DE \parallel AB$, but D and E are not mid-points of b, a.

Proposition III.

282. Theorem. *Triangles, or parallelograms, which have an angle in one equal to an angle in the other, have the same ratio as the products of the including sides.*

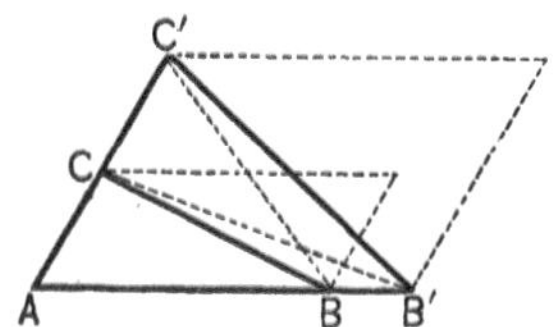

Given two triangles ABC, $AB'C'$, having an angle, A, of one equal to an angle, A, of the other.

To prove that $\triangle ABC : \triangle AB'C' = AB \cdot AC : AB' \cdot AC'$.

Proof. 1. Suppose them placed with $\angle A$ in common; draw BC', $B'C$.

Then $\triangle ABC : \triangle AB'C = AB : AB'$.   Why?

2. And $\triangle AB'C : \triangle AB'C' = AC : AC'$,

their bases being AC, AC'.   Why?

3. $\therefore \triangle ABC : \triangle AB'C' = AB \cdot AC : AB' \cdot AC'$.   Ax. 6

4. And $\because$ the $\square$ in the figure are double the $\triangle$, the theorem is true for parallelograms.

Corollary. *Similar triangles have the same ratio as the squares of their corresponding sides.*

For if the $\triangle$ are similar, $BC \parallel B'C'$, and the ratio $AB : AB'$ equals, and may be substituted for, the ratio $AC : AC'$, thus making the second member of step 3, $AB^2 : AB'^2$.

Exercises. 428. Prove prop. III, changed to read, "an angle in one *supplemental* to an angle in the other."

429. Prove the converse of prop. III, cor. : If two triangles have the same ratio as the squares of any two corresponding sides, they are similar.

Proposition IV.

283. Theorem. *Similar polygons have the same ratio as the squares of their corresponding sides.*

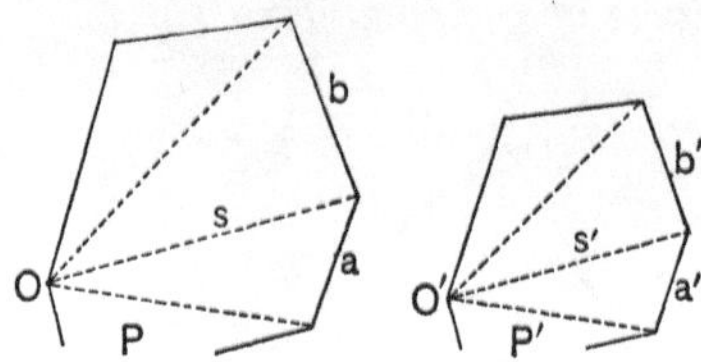

Given P and P', two similar polygons; sides a, b, ..., corresponding to sides a', b', ...; diagonal s corresponding to diagonal s'.

To prove that $P : P' = a^2 : a'^2$.

Proof. 1. Suppose P and P' divided into similar $\triangle$ of bases a and a', b and b', ..., by diagonals from corresponding points O, O'.   IV, prop. XXI, cor. 4

Then $\triangle\, Oa : \triangle\, O'a' = a^2 : a'^2$,   Prop. III, cor.

and $\triangle\, Oa : \triangle\, O'a' = s^2 : s'^2 = \triangle\, Ob : \triangle\, O'b' = \cdots$

Why ?

2. $\therefore \triangle\, Oa + Ob + \cdots : \triangle\, O'a' + O'b' + \cdots = \triangle\, Oa : \triangle\, O'a'$.

IV, prop. VI

3. $\because \triangle\, Oa + Ob + \cdots = P$, and $\triangle\, O'a' + O'b' + \cdots = P'$,

$\therefore P : P' = a^2 : a'^2$.

Exercises.   430. If the vertices, A, B, C, of a triangle are joined to a point O within the triangle, and if $A\,O$ produced cuts a at D, then

$$\triangle\, ABO : \triangle\, AOC = BD : DC.$$

431. If two triangles are on equal bases and between the same parallels, then any line parallel to their bases, cutting the triangles, will cut off equal triangles.

432. Two equilateral triangles have their areas in the ratio of $1 : 2$. Find the ratio of their sides to the nearest 0.01.

2. PARTITION OF THE PERIGON.

PROPOSITION V.

284. Problem. *To bisect a perigon.*

Construction and Proof. A special case under I, prop. XXVIII.

COROLLARY. *A perigon can be divided into 2^n equal angles.*

PROPOSITION VI.

285. Problem. *To trisect a perigon.*

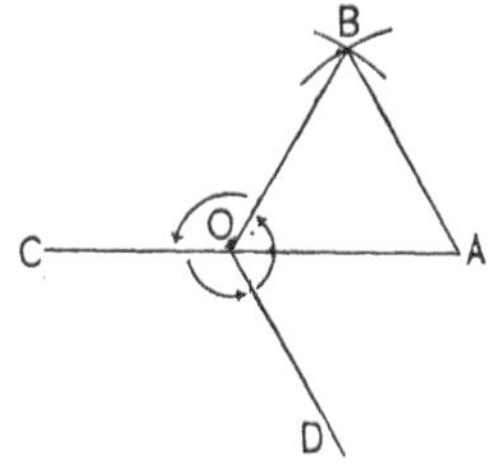

Given the perigon with vertex O.

Required to trisect it.

Construction. 1. On any line OA, from O, construct an equilateral $\triangle OAB$. Authority ?

2. Produce AO to C, and bisect $\angle COB$ by OD.
Then the perigon is trisected by OB, OC, OD.

Proof. 1. $\because \angle AOB = 60°$, I, prop. XIX, cor. 8
$\therefore \angle BOC$, supplement of $\angle AOB = 120°$.

2. $\therefore \angle COB$, conjugate of $\angle BOC = 240°$.

3. $\therefore \angle COD = \angle DOB = 120°$. Const. 2

Corollary. *A perigon can be divided into $3 \cdot 2^n$ equal angles.*

For if $n = 0$, then $3 \cdot 2^n = 3 \cdot 1 = 3$, so that the corollary reduces to the problem itself. If $n = 1$, then $3 \cdot 2^n = 6$, and by bisecting $\angle BOC$, COD, DOB, the perigon is divided into 6 equal angles. Similarly, by bisecting again, the perigon is divided into $3 \cdot 2^2 = 12$ equal angles, and so on.

Proposition VII.

286. **Problem.** *To divide a perigon into five equal angles.*

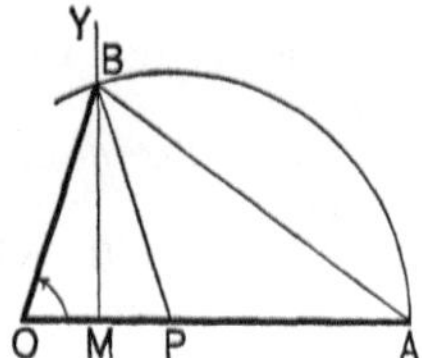

Given the perigon with vertex O.

Required to divide it into five equal angles.

Construction. 1. Draw OA, and divide it at P so that $OP \cdot OA = PA^2$.   IV, prop. XXVI

2. Draw MY, the $\perp$ bisector of OP.   I, prop. XXXI

3. With center P and radius PA describe an arc cutting MY in B.   § 109

4.   Draw OB.

Then $\angle AOB = \frac{1}{5}$ of a perigon.

Proof. 1.   Draw AB and PB.

Then $\because OP \cdot OA = PA^2$,

and $\because OA > PA$,

$\therefore OP < PA$, and $\therefore MP < PA$.

$\therefore \overset{\frown}{AB}$ cuts MY.

2. $\because \triangle OPB$ is isosceles, $OB = PB = PA$, the radius.

3. Also, $OA^2 + OB^2 = AB^2 + 2\,OM \cdot OA.$

II, prop. IX, cor. 1

And $\qquad \because 2\,OM = OP,$

$\qquad \therefore OA^2 + OB^2 = AB^2 + OP \cdot OA.$

4. And $\quad \because OB^2 = PA^2 = OP \cdot OA,$

$\qquad \therefore OA^2 + OB^2 = AB^2 + OB^2$, from step 3,

$\qquad\qquad \therefore OA^2 = AB^2,$

and $\qquad \therefore OA = AB.$

5. $\therefore \angle OBA = \angle O = \angle BPO,$ $\qquad$ I, prop. III

$\qquad\qquad\qquad = \angle A + \angle PBA.$ $\quad$ I, prop. XIX

And $\qquad \because \angle A = \angle PBA,$ $\qquad\qquad$ I, prop. III

$\qquad\qquad \therefore \angle O = 2 \angle A.$

6. $\therefore \angle O$ is $\frac{2}{5}$ of a st. $\angle$, or $\frac{1}{5}$ of a perigon.

I, prop. XIX

Corollary. *A perigon can be divided into $5 \cdot 2^n$ equal parts.*

For if $n = 0$, then $5 \cdot 2^n = 5 \cdot 1 = 5$, so that the corollary reduces to the problem itself. If $n = 1$, then $5 \cdot 2^n = 5 \cdot 2 = 10$, and by bisecting $\angle AOB$, the resulting angle is $\frac{1}{10}$ of a perigon. Similarly, by bisecting again, $\frac{1}{20}$ of a perigon is formed, and so on.

Exercises. 433. In the figure on p. 206, let $OP = x$, $PA = r$; then show that $x = \dfrac{r}{2}(\sqrt{5} - 1)$. (Omit exs. 433, 434 if the student has not had quadratic equations.)

434. In the same figure, if $OP = x$ and $OA = a$, show that $x = \dfrac{a}{2}(3 - \sqrt{5})$.

435. On the sides a, b, c, of an equilateral triangle, points X, Y, Z are so taken that $BX : XC = CY : YA = AZ : ZB = 2 : 1$. Find the ratio of $\triangle XYZ$ to $\triangle ABC$.

PROPOSITION VIII.

287. Problem. *To divide a perigon into fifteen equal angles.*

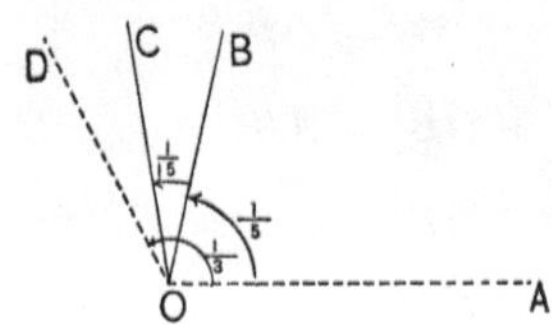

Solution.   1. Make $\angle AOB = \frac{1}{6}$ of a perigon.   Prop. VII

   2. Make $\angle AOD = \frac{1}{3}$ of a perigon.   Prop. VI

   3. Bisect $\angle BOD$.   I, prop. XXVIII

   4. Then $\angle BOC = \frac{1}{2}(\frac{1}{3} - \frac{1}{6})$ perigon $= \frac{1}{15}$ of a perigon.

COROLLARY. *A perigon can be divided into $15 \cdot 2^n$ equal angles.*

Explain.

288. NOTE. That a perigon could be divided into 2^n, $3 \cdot 2^n$, $5 \cdot 2^n$, $15 \cdot 2^n$ equal angles, was known as early as Euclid's time. By the use of the compasses and straight edge no other partitions were deemed possible. In 1796 Gauss found, and published the fact in 1801, that a perigon could be divided into 17, and hence into $17 \cdot 2^n$ equal angles; furthermore, that it could be divided into $2^m + 1$ equal angles if $2^m + 1$ was a prime number; and, in general, that it could be divided into a number of equal angles represented by the product of different prime numbers of the form $2^m + 1$. Hence it follows that a perigon can be divided into a number of equal angles represented by the product of 2^n and one or more different prime numbers of the form $2^m + 1$. It is shown in the Theory of Numbers that if $2^m + 1$ is prime, m must equal 2^p; hence the general form for the prime numbers mentioned is $2^{2^p} + 1$. Gauss's proof is only semi-geometric, and is not adapted to elementary geometry.

Exercises.   436. Including the divisions of a perigon suggested by Gauss, there are 25 possible divisions below 100.   What are they?

437. As in ex. 436, there are 13 possible divisions between 100 and 300.   What are they?

3. REGULAR POLYGONS.

Proposition IX.

289. **Problem.** *To inscribe in a circle a regular polygon having a given number of sides.*

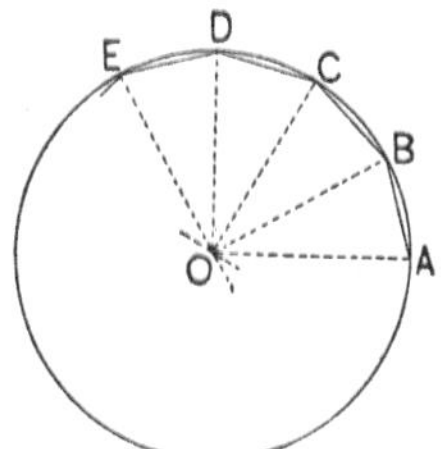

Given   a circle with center O and radius OA.

Required   to inscribe in the circle a regular n-gon.

Construction. 1. Divide the perigon O into n equal parts (n being limited as in props. V–VIII and cors.) as AOB, BOC, COD, ..., B, C, D, ... lying on the circumference.

    2. Draw AB, BC, CD .....   § 28

    Then $ABCD$ ..... is an inscribed regular n-gon.

Proof. 1. $\triangle AOB \cong \triangle BOC \cong \triangle COD \cong$ .....,

    and $AB = BC = CD =$ .....   I, prop. I

    2. $\therefore \overset{\frown}{AB} = \overset{\frown}{BC} = \overset{\frown}{CD} =$ .....   III, prop. IV

    3. $\therefore \angle DCB = \angle CBA =$ .....,

    $\because$ each stands on $(n-2)$ arcs equal to $\overset{\frown}{AB}$.

                    III, prop. XI, cor. 1

    4. $\therefore ABCD$ ..... is an inscribed regular polygon.

                    §§ 92, 201

Cᴏʀᴏʟʟᴀʀɪᴇs.   1. *The side of an inscribed regular hexagon equals the radius of the circle.*

Then $\angle AOB = \frac{1}{6}$ of $360° = 60°$; $\therefore \angle BAO$, which $= \angle OBA = 60°$. $\therefore \triangle ABO$ is equilateral.

2. *An inscribed equilateral polygon is regular.*

For by step 3 of the proof it is also equiangular; and being both equilateral and equiangular, it is regular.

Pʀᴏᴘᴏsɪᴛɪᴏɴ X.

290. Problem.   *To circumscribe about a circle a regular polygon having a given number of sides.*

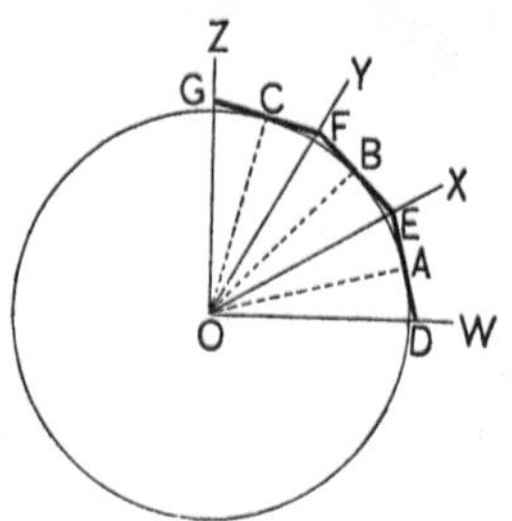

Given   a circle with center O and radius OA.

Required   to circumscribe about this circle a regular n-gon.

Construction.   1. Divide the perigon O into n equal parts (n being limited as in props. V–VIII and cors.) by lines OW, OX, OY, ...

2. Bisect $\angle s$ WOX, XOY, ... by radii to A, B, ...
I, prop. XXVIII

3. From A, B, C, ... draw tangents to meet OW at D, OX at E, ...   III, prop. XXVI

Then $DEFG$ ..... is the required polygon.

Proof. 1. $\because DE \perp OA$, $EF \perp OB$, ...,   III, prop. IX, cor. 2

$\therefore \triangle OAE \cong \triangle OBE$, and $AE = BE$, $OE = OE$.

I, prop. II

2. $\therefore$ the tangents from A and B meet OX at the same point, E.

3. And $\because \angle DOE = \angle EOF$,   Const. 1

and $\angle OED = \angle FEO$,   Step 1

$\therefore \triangle DOE \cong \triangle EOF$, and $DE = EF$.   Why ?

4. Also, $\because \angle GFE = \angle FED$, each being the supplement of an $\angle$ equal to $\angle WOX$,   (Name the $\triangle$s.)

I, prop. XXI, cor.

$\therefore DEFG$ ..... is a circumscribed regular polygon.

§§ 92, 201

Corollaries. 1. *The side of a regular hexagon circumscribed about a circle of diameter* 1, *is* $1/\sqrt{3}$, *or* $\frac{1}{3}\sqrt{3}$.

For it is (as in prop. IX, cor. 1) the side of an equilateral $\triangle$ whose altitude is $\frac{1}{2}$. This is easily shown to be $1/\sqrt{3}$. (Show it.)

2. *A circumscribed equiangular polygon is regular.*

Prove that any two adjacent sides are equal.

Exercises. 438. In a right-angled triangle, any polygon on the hypotenuse equals the sum of two similar polygons described on the sides as corresponding sides of those polygons.

(Suggestion : $P_2 : P_3 = b^2 : c^2$;

$\therefore P_2 + P_3 : P_3 = b^2 + c^2 : c^2$
$\qquad\qquad = a^2 : c^2 = P_1 : P_3$;

$\therefore P_2 + P_3 : P_1 = P_3 : P_3 = 1$. This is one of the generalized forms of the Pythagorean theorem.)

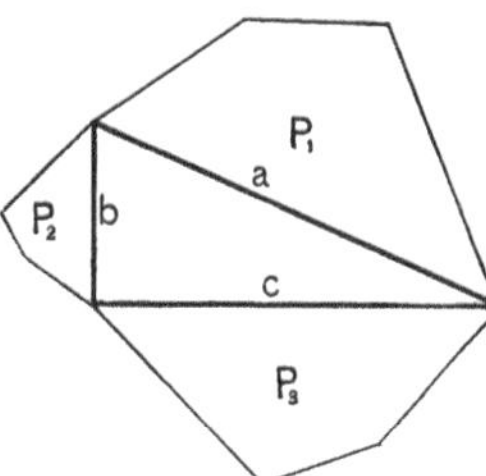

439. If r is the radius of the circle, and s is the side of the inscribed equilateral triangle, then $s = r\sqrt{3}$.

Proposition XI.

291. Problem. *To circumscribe a circle about a given regular polygon.*

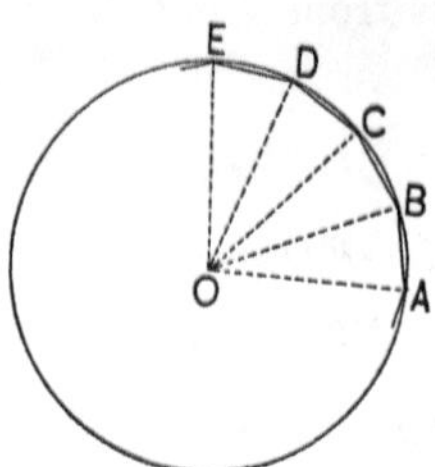

Given the regular polygon $ABCD$ .....

Required to circumscribe a circle about it.

Construction. Bisect $\angle s$ DCB, CBA, the bisectors meeting at O. Then O is the center and OB the radius.

Proof. 1. Draw OA, OD, OE, ... Then $\because$ $\angle s$ OCB, CBO are halves of oblique $\angle s$, each is less than a rt. $\angle$.

2. $\therefore$ CO and BO cannot be $\parallel$, and they meet as at O.

3. And $\because$ $\angle CBO = \angle OBA$,   Const.
and $AB = BC$,   § 92, def. reg. pol.

4. $\therefore \triangle ABO \cong \triangle CBO$, and $OA = OC$.   Why ?
Similarly each of the lines OB, OD, ... $= OC$.

5. $\therefore$ O is the center, and OA, OB, ... are radii. § 108

292. Nᴏᴛᴇ. The inscription and circumscription of regular polygons are seen to depend upon the partition of the perigon. Elementary geometry is thus limited to the inscription and circumscription of regular polygons of 2^n, $3 \cdot 2^n$, $5 \cdot 2^n$, $15 \cdot 2^n$ sides ; or, since the discovery by Gauss, to polygons the number of whose sides is represented by the product of 2^n and one or more different prime numbers of the form $2^m + 1$.

In addition to regular convex polygons, cross polygons can also be regular, the common five-pointed star being an example.

Proposition XII.

293. **Problem.** *To inscribe a circle in a given regular polygon.*

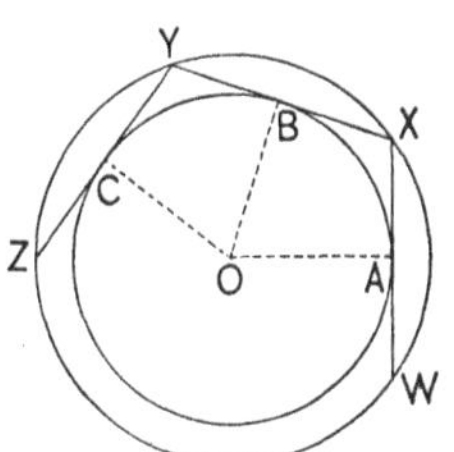

Given   a regular polygon WXY .....

Required   to inscribe a circle in it.

Construction. 1. Circumscribe a circle about it.   Prop. XI

2. From center O of this $\odot$ draw $OA \perp WX$.

I, prop. XXX

With center O, and radius OA, a $\odot$ may be inscribed.

Proof. 1. Draw OB, OC, ... $\perp XY$, YZ, ...

Then $\because$ OA bisects WX, $\therefore$ A lies between W and X, and so for B, C, ...   III, prop. V

2. And $\because$ $WX = XY = $ ....., $\therefore$ $OA = OB = $ .....

III, prop. VII

3. $\therefore$ if with center O and radius OA a $\odot$ is described, then WX, XY, ... will be tangent to the $\odot$,

III, prop. IX, cor. 3

and $\therefore$ the $\odot$ is inscribed in the polygon.

§ 201, def. inscr. $\odot$

Exercises. **440.** Solve prop. XI by bisecting the sides AB, BC by perpendiculars, thus determining O.

441. Inscribe a regular cross pentagon in a circle. (The regular cross pentagon, the pentagram, was the badge of the Pythagorean school.)

COROLLARIES.   1. *The inscribed and circumscribed circles of a regular polygon are concentric.*

For from step 2 of the construction and step 2 of the proof, O is the center of both circles.

2. *The bisectors of the angles of a regular polygon meet in the common in- and circumcenter.*

For by the proof of prop. XI they meet in O, and by cor. 1 O is the common in- and circumcenter.

3. *The perpendicular bisectors of the sides of a regular polygon meet in the common in- and circumcenter.*   (Why?)

294. Definitions. The radius of the circumscribed circle is called the **radius** of a regular polygon; the radius of the inscribed circle, the **apothem** of that polygon; the common center of the two circles, the **center** of that polygon.

E.g. in the figure below, r is the radius, m the apothem, and O the center of the regular polygon, part of which is shown as inscribed in the circle.

PROPOSITION XIII.

295. Theorem. *The area of a regular polygon equals half the product of the apothem and perimeter.*

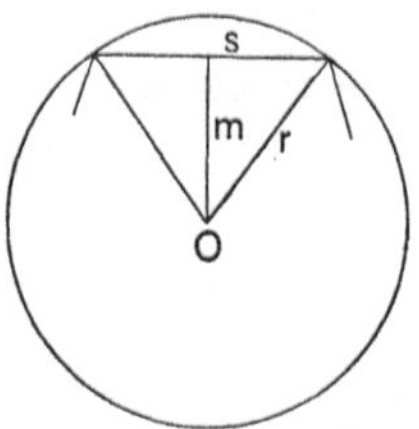

Given   an inscribed regular polygon, or area a, perimeter
            p, apothem m.

To prove that $a = \tfrac{1}{2}\, mp.$

Proof.   Let O be the center and r the radius of the circumscribed circle.

Let t be one of the $\triangle$ formed by joining O to two consecutive vertices, and s a side of the polygon.

Then area t equals $\frac{1}{2}\,ms$.   Why ?

$\therefore$ the area of the polygon equals the sum of the areas of the triangles $= \frac{1}{2}\,m \times$ the sum of the sides $= \frac{1}{2}\,mp$.   Ax. 2

Corollaries.   1. *The areas of regular polygons of the same number of sides are proportional to the squares of their apothems, of their radii, or of their sides.*

For $\dfrac{a}{a'} = \dfrac{\frac{1}{2}mp}{\frac{1}{2}m'p'} = \dfrac{mp}{m'p'}$; and from similar $\triangle$ and IV, prop. XX,

$\dfrac{m}{m'} = \dfrac{r}{r'} = \dfrac{s}{s'} = \dfrac{p}{p'}$; $\therefore$ by substitution $\dfrac{a}{a'} = \dfrac{m^2}{m'^2} = \dfrac{r^2}{r'^2} = \dfrac{s^2}{s'^2}$.

2. *The perimeters of regular polygons of the same number of sides are proportional to their apothems, their radii, or their sides.*

Proved with cor. 1. ___________

Exercises.   **442.** The distance from the center to a side of the inscribed equilateral triangle equals $r/2$.

443. Draw a diameter AB of a circle with center O; then with center A and radius AO draw an arc cutting the circumference in C, D; draw CD, DB, BC, and prove $\triangle\,BCD$ equilateral.

444. The area of an inscribed regular hexagon is a mean proportional between the areas of the inscribed and circumscribed equilateral triangles.

445. Show how, with compasses alone, to divide a circumference into six equal arcs.

446. Prove that if AB, CD, two diameters of a circle, are perpendicular to each other, then $ACBD$ is an inscribed square.

447. Let OX be the perpendicular bisector of line-segment AB at O; lay off on OX, $OD = AO$; and, on DX, lay off $DC = DB$; then prove that C is the center of the $\odot$ circumscribed about the regular octagon of which AB is a side.

4. THE MENSURATION OF THE CIRCLE.

296. Postulate of Limits. *The circle and its circumference are the respective limits which the inscribed and circumscribed regular polygons and their perimeters approach, if the number of their sides increases indefinitely.*

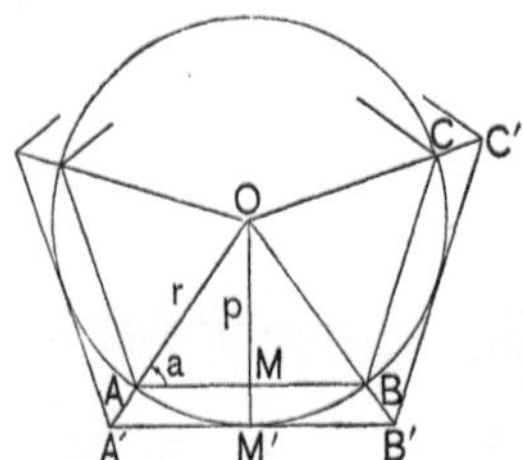

The following may be read by the student in connection with the postulate, although it does not constitute a proof:

1. In the figure, suppose an in- and circumscribed regular n-gon represented. Then each exterior angle equals $\dfrac{360°}{n}$ in each figure.

2. $\therefore$ each interior angle equals $180° - \dfrac{360°}{n}$, and $\therefore \angle a = 90° - \dfrac{180°}{n}$.

3. $\therefore$ if n increases indefinitely, $\angle a \doteq 90°$, and $p \doteq r$.

4. $\therefore$ the inscribed polygon $\doteq$ the circle, and its perimeter $\doteq$ the circumference. Similarly for the circumscribed polygon.

Corollaries. 1. *The circumscribed regular polygon and its perimeter are respectively greater than the circle and its circumference ; the inscribed, and its perimeter, less.*

2. *If, on any finite closed curve, n points are assumed equidistant from each other, and each connected with the succeeding point by a straight line, then the curve is the limit which the broken line approaches if n increases indefinitely.*

Proposition XIV.

297. Theorem. *The ratio of the circumference to the diameter of a circle is constant.*

Proof. 1. Suppose any two circles, of circumferences c, c', radii r, r', and diameters d, d', respectively, to have similar regular polygons inscribed in them, of perimeters p, p', respectively.

Then $p : p' = r : r'$,   Prop. XIII, cor. 2

$\qquad = 2\,r : 2\,r' = d : d'.$ IV, prop. VIII

2. And $\because r$, r', d, d' do not change when the number of sides of the polygons is doubled, quadrupled, ...,

$\qquad\qquad\qquad\qquad$ § 294, def. radius polyg.

and $\because p \doteq c$, and $p' \doteq c'$,   § 296, post. of limits

$\therefore c : c' = d : d'.$ IV, prop. IX

3. $\quad \therefore c : d = c' : d' =$ the same for any ⓢ. IV, prop. III

Note. This constant ratio $c : d$ is designated by the symbol π (pi), the initial letter of the Greek word for circumference (*periphereia*). The value of π is discussed in prop. XVII.

Corollaries. 1. $c = \pi\mathrm{d}$, *or* $2\,\pi\mathrm{r}$.

For if $\dfrac{c}{d} = \pi$, then $c = \pi d$.

2. *If the radius of a circle is* 1, *then* $c = 2\,\pi$, *or a semi-circumference equals* π.

3. *The circumferences of two circles are proportional to their radii.*

For $\dfrac{c}{c'} = \dfrac{2\,\pi r}{2\,\pi r'} = \dfrac{r}{r'}.$

Exercises. 448. Find, in terms of the radius of the circle, r, the side, apothem, and area of the inscribed and circumscribed equilateral triangle.

449. Also of the inscribed and circumscribed square.

PROPOSITION XV.

298. Problem. *Given the sides of the regular inscribed and circumscribed n-gons, to find the side of the regular circumscribed 2 n-gon.*

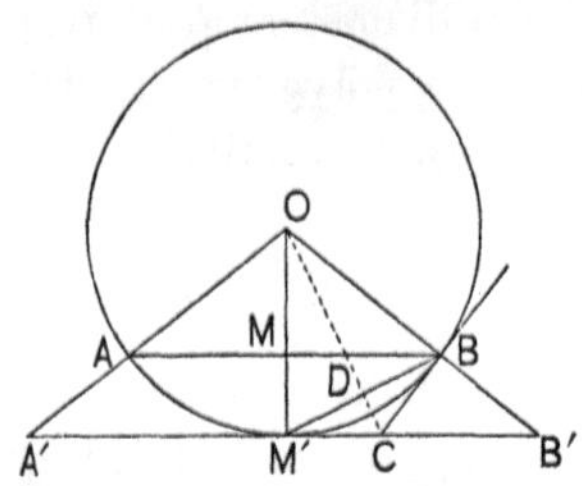

Solution. 1. In the figure,

let AB = a side of the regular inscribed n-gon, i_n;

" $A'B'$ = " " " circumscribed " c_n.

Then BM'

= " " " inscribed $2 n$-gon, i_{2n};

and BC

= $\frac{1}{2}$ " " " circumscribed " c_{2n}.

2. But $\because$ OC bisects $\angle M'OB'$,   Why?

$$\therefore CB' : M'C = OB' : OM' \;(= OB),$$

IV, prop. XIII

$$= A'B' : AB, \quad \text{IV, prop. X, cor. 3}$$

$$= c_n : i_n.$$

3. $\therefore CB' + M'C : M'C = c_n + i_n : i_n,$ IV, prop. V

or $\quad M'B' : M'C = c_n + i_n : i_n.$

4. $\therefore \quad 2 M'B' : 2 M'C = c_n + i_n : i_n.$ IV, prop. VIII

5. $\therefore \quad c_n : c_{2n} = c_n + i_n : i_n,$

or $\quad\quad c_{2n} = \dfrac{c_n \cdot i_n}{c_n + i_n}.$

Proposition XVI.

299. Problem. *Given the sides of the regular inscribed n-gon and the regular circumscribed 2 n-gon, to find the side of the regular inscribed 2 n-gon.*

Solution. 1. In the figure on p. 218,

$$\triangle M'BM \backsim \triangle CM'D. \qquad \text{Why ?}$$

$$\therefore M'B : BM = CM' : M'D. \qquad \text{Why ?}$$

2. Or $\qquad i_{2n} : \tfrac{1}{2} i_n = \tfrac{1}{2} c_{2n} : \tfrac{1}{2} i_{2n}.$ Why ?

3. $\qquad \therefore i_{2n} = \tfrac{1}{2} \sqrt{2\, c_{2n} \cdot i_n}.$ Why ?

Corollaries. 1. *If* p_n, p_{2n}, P_n, P_{2n} *represent the perimeters of the polygons with sides* $i_n, i_{2n}, c_n, c_{2n},$ *respectively, then*

$$(1)\ \ P_{2n} = \frac{2\, P_n \cdot p_n}{P_n + p_n}, \ \ and \ \ (2)\ \ p_{2n} = \sqrt{p_n \cdot P_{2n}}.$$

For $c_{2n} = P_{2n}/2\,n$, $c_n = P_n/n$, $i_{2n} = p_{2n}/2\,n$, and $i_n = p_n/n$; substitute these in the final steps of props. XV, XVI. From prop. XV,

$$\frac{P_{2n}}{2\,n} = \frac{P_n/n \cdot p_n/n}{P_n/n + p_n/n} = \frac{P_n \cdot p_n/n}{P_n + p_n}.$$

$$\therefore P_{2n} = \frac{2\, P_n \cdot p_n}{P_n + p_n}.$$

From prop. XVI,

$$\frac{p_{2n}}{2\,n} = \frac{1}{2} \sqrt{2 \cdot \frac{P_{2n}}{2\,n} \cdot \frac{p_n}{n}}.$$

$$\therefore p_{2n} = \sqrt{p_n \cdot P_{2n}}.$$

2. $c_n = \dfrac{r \cdot i_n}{\sqrt{r^2 - \tfrac{1}{4} i_n^{\,2}}},$ *where* r *is the radius.*

For $c_n : i_n = r : OM = r : \sqrt{r^2 - \left(\dfrac{i_n}{2}\right)^2} = r : \sqrt{r^2 - \tfrac{1}{4} i_n^2}.$

$$\therefore c_n = \frac{r \cdot i_n}{\sqrt{r^2 - \tfrac{1}{4} i_n^{\,2}}}, \ \ \text{by multiplying by } i_n.$$

Proposition XVII.

300. Theorem. *The approximate value of π is 3.14159 +.*

Proof. 1. In a regular hexagon inscribed in a circle of diameter 1, $i_6 = \frac{1}{2}$, and $\therefore p_6 = 3$. Prop. IX, cor. 1

2. Of the regular hexagon circumscribed about that $\odot$, $c_6 = 1/\sqrt{3}$. Prop. X, cor. 1

3. $\therefore P_6 = 6 \cdot c_6 = 3.4641016 \ldots$

4. From p_6 and P_6 can be found p_{12} and P_{12}.

Props. XV, XVI

5. From p_{12} and P_{12} can be found p_{24} and P_{24}, and so on. Props. XV, XVI

If the process were continued to a 1536-gon, p_{1536} would be found to be 3.1415904, and P_{1536} would be found to be 3.1415970.

6. And $\therefore c$, or $\pi \cdot d$, which equals $\pi \cdot 1$ or π, lies between p_n and P_n, however large n may be,

§ 296, post. of limits, cor. 1

$\therefore \pi$ lies between 3.1415904 and 3.1415970, and is, therefore, approximately 3.14159 +.

Exercises. **450.** The diagonals of a regular pentagon cut each other in extreme and mean ratio.

451. If $ABCDE$ is a regular pentagon, and AD cuts BE at P, prove that $AP : AE = AE : AD$.

452. To construct a regular pentagon equal to the sum of two given regular pentagons.

453. Find, in terms of the radius of the circle, r, the side of the inscribed regular pentagon. (Omit unless ex. 433 was taken.)

454. Also of the inscribed and circumscribed regular hexagon.

455. Also of the inscribed and circumscribed regular dodecagon.

456. Also of the inscribed regular decagon. (Depends on ex. 453.)

301. Notes. The computation in prop. XVII, which the student is not expected to make, is as follows :

No. of sides	p	P
6	3.	3.4641016
12	3.1058285	3.2153903
24	3.1326286	3.1596599
48	3.1393502	3.1460862
96	3.1410319	3.1427146
192	3.1414524	3.1418730
384	3.1415576	3.1416627
768	3.1415838	3.1416101
1536	3.1415904	3.1415970
. . .	. . .	. . .

302. The following historical notes on π are inserted to show the student how the subject of the mensuration of the circle has grown.

The early approximation for π, in use among the ancient people, was 3. See I Kings, vii, 23 ; II Chron. iv, 2. " What is three hand-breadths around is one hand-breadth through." — *The Talmud.*

Ahmes, however, gave the equivalent of 3.1604.

Archimedes seems to have been the first to employ geometric methods similar to that of props. XV, XVI for approximating π. He announced, " The circumference of a circle exceeds 3 times the diameter by a part which is less than $\frac{1}{7}$, but more than $\frac{10}{71}$, of the diameter."

Hero of Alexandria used both 3 and $3\frac{1}{7}$.

Ptolemy of Alexandria gave $3\frac{17}{120}$.

Aryabhatta found 3.1416, by a method similar to that of prop. XVII.

Brahmagupta used the values of Archimedes ; also $\frac{3927}{1250}$ and $\frac{754}{240}$, the last being only another form for Ptolemy's.

Metius gave the easily remembered value 355/113.

Ludolph van Ceulen computed π to the equivalent of over 30 decimal places (the decimal fraction was not yet invented), and wished it engraved on his tomb at Leyden. On this account π is often called in Germany, " the Ludolphian number."

Vega carried it to 140 decimal places.

Dase carried it to 200 decimal places.

Richter carried it to 500 decimal places. More recently Shanks carried it to 707 decimal places.

The symbol π is first used in this sense in Jones's " Synopsis Palmariorum Matheseos," London, 1706.

303. Definition. It is now necessary to extend our idea of **equal surfaces.** The definition at the beginning of Book II, § 142, is true, and it suffices for the cases there under consideration. But when curvilinear figures are compared with rectilinear, it is impossible to cut the surfaces into parts respectively congruent. Hence, we enlarge the definition, thus: *Two surfaces are said to be equal if they have the same numerical measure in terms of a common unit.*

Thus, a circle having an area of $2\,m^2$ would equal a rectangle $2\,m$ long by $1\,m$ broad, even though they could not be cut into parts respectively congruent.

304. Table of Values. The following table of values of expressions involving π will be found useful in computations concerning the circle, sphere, cylinder, cone, etc.:

$$\pi = 3.14159 \qquad \sqrt{\pi} = 1.77245 \qquad 180°/\pi = 57°.29578$$
$$\pi/4 = 0.78540 \qquad 1/\sqrt{\pi} = 0.56419 \qquad \pi/180 = 0\ .01745$$
$$1/\pi = 0.31831 \qquad \pi\sqrt{2} = 4.44288 \qquad \text{Approximate values:}$$
$$\pi^2 = 9.86960 \qquad \sqrt{\pi/2} = 1.25331 \qquad \pi = \tfrac{22}{7} = 3\tfrac{1}{7},\ \tfrac{355}{113}.$$

The table is repeated, with other tables of value in numerical computations, at the end of this work.

305. Radian Measure of Angles and Arcs. Since if $A =$ any central angle and $a =$ its arc,

$$A : \text{st.} \angle \quad = a : \text{semicircumf.} = a : \pi r.$$
$$\therefore A : \text{st.} \angle/\pi = a : r,$$
$$\text{or } A : 180°/\pi = a : r,$$
$$\text{or } A : 57°.29 + = a : r.$$

That is, the ratio of a central angle to st. $\angle/\pi$ equals the ratio of its arc to an arc of the same length as the radius. Just as the "degree" is the unit for both angle and arc measure, it being understood to be $\frac{1}{360}$ of a perigon in the one case and $\frac{1}{360}$ of a circumference in the other, so a special name is given to st. $\angle/\pi$ and to an arc which equals a radius in length; this name is **radian.** In other words, a radian is $\dfrac{1}{\pi}$ of a st. $\angle$, in angle measure, and $\dfrac{1}{\pi}$ of a semicircumference, or an arc equal to a radius in length, in arc measure.

Since $r = \dfrac{1}{\pi}\,180°$, $\therefore r = 57°.29+$, where r stands for radian.

Also $\because 180° = \pi r$,　$\therefore 1° = \dfrac{\pi r}{180}$ or $\dfrac{\pi}{180}$ of a radian, or .0174533 of a radian.

In most work in advanced mathematics the radian measure is used exclusively. In common measurements the degree is used. It is necessary in this work to use both.

It is customary to express an angle in radians by the Greek letters a (alpha), β (beta), γ (gamma), ..., the first letters of that alphabet.

306. Corollary. *The length of an arc equals the product of the radius by the angle in radians.*

For if $a = $ length of arc, and $\alpha = $ its $\angle$ in radians, then $\dfrac{a}{c} = \dfrac{\alpha}{2\,\pi}$, $\therefore a = \alpha \cdot \dfrac{2\,\pi r}{2\,\pi} = \alpha \cdot r.$

Exercises.　457. Express the following in radians: 10°, 21° 20′, 57°, 58°, 90°.

458. Express the following in degrees: 1.3090 r, .8058 r, .3636 r, .1687 r, .0029 r.

459. Express the following in radians: 100°, 180°, 270°.

460. Express the following in degrees: 3.4907 r, 5.2359 r, 6.2832 r, πr.

461. Find the lengths of arcs of 47° 50′, 61° 20′, 75° 40′, the radius being 10.

462. Given the lengths of the following arcs, to find the radii of the various circles: 75° 10′, 131.19; 32° 20′, 2.822; 4°, .0698.

463. Show that the perimeters of the inscribed and circumscribed squares, the diameter of the circle being 1, are respectively 2.8284271 and 4; hence, find the perimeters of the inscribed and circumscribed regular octagons, and thus show that the value of π may be approximated in this way.

464. The circumferences of certain Ⓢ are 43.9823, 84.8230, 128.8053, 185.5340, 204.2035; find the diameters.

Proposition XVIII.

307. Theorem. *The area of a circle equals half the product of its circumference and radius.*

Given a, c, r, the area, circumference, and radius of a circle.

To prove that $\qquad\qquad a = \tfrac{1}{2} cr.$

Proof. 1. If a', p represent the area and perimeter of a circumscribed regular polygon, then the apothem of that polygon is r. $\qquad$ § 294

    2. And $\;a' = \tfrac{1}{2} pr.$ $\qquad\qquad\qquad$ Prop. XIII

    3. But $\;a' \doteq a$, and $\tfrac{1}{2} pr \doteq \tfrac{1}{2} cr.$ $\quad$ § 296, post. of limits

    4. $\qquad \therefore a = \tfrac{1}{2} cr.$ $\qquad\qquad$ IV, prop. IX, cor. 1

Corollaries.   1. $a = \pi r^2.$

For $c = 2 \pi r.$

2. $a = \dfrac{c^2}{4 \pi}.$ (Why ?)

3. *If* s *represents the area of a sector, and* α *its angle in radians, then* s $= r^2 \alpha / 2.$

For $s : \pi r^2 = \alpha : 2 \pi.$ (IV, prop. XVI, cor.)

4. *Of two unequal circles, the greater has the greater circumference.*

For, by cor. 2, $\qquad\qquad a = \dfrac{c^2}{4 \pi}.$

$$\therefore 4 \pi a = c^2.$$

$\therefore$ as the area increases, the circumference increases also.

5. *The areas of two circles are proportional to the squares of their radii.*

For $\dfrac{a}{a'} = \dfrac{\pi r^2}{\pi r'^2} = \dfrac{r^2}{r'^2}$

308. Historical Note on Quadrature of the Circle. The expression, "to square the circle," means to find the side of a square whose area equals that of a given circle. The solution of this problem by elementary geometry has been proved to be impossible. It nevertheless occupied the attention of many mathematicians before this impossibility was shown, and many ignorant people still attempt it. Some of the Pythagorean school claimed to have solved it, Anaxagoras (died 428 b.c.) wrote upon it, and hundreds of writers since then have discussed the subject. It is closely related to finding a straight line equal to a given circumference ("to rectify the circumference"), and the two depend upon finding the value of π exactly. That π cannot be expressed exactly, nor as the root of a rational algebraic equation, was shown by Lindemann in 1882.

For the mathematical discussion, see Klein's "Famous Problems of Elementary Geometry," translated by the authors. (Boston, Ginn & Co.)

Exercises.   465. What is the radius of that circle of which the number of square units of area equals the number of linear units of circumference?

466. Also, of which the number of square units of area equals the number of linear units of radius?

467. Give a formula for a in terms of d, and the constant π.

468. A circle equals a triangle of which the base equals the circumference and the altitude equals the radius.

469. Find the areas of circles with radii 5, 7, 21, 35, 47, 50. (In these computations, for uniformity let $\pi = 3.1416$.)

470. Also with diameters 2, 8, 11, 31, 42, 97.

471. Find the radii of circles of areas 78.5398, 2042.8206, 4536.4598.

472. Also the diameters of circles of areas 2123.7166, 3318.3072, 56.745017.

473. Also the circumferences of circles of areas 95.0332, 452.3893.

474. Also the areas of circles of circumferences 267.0354, 191.6372.

475. The area of the ring formed between the circumferences of two concentric circles of radii r_1, r_2, where $r_1 > r_2$, is $\pi (r_1 + r_2)(r_1 - r_2)$.

476. The area of that portion of the ring of ex. 475 cut off by the arms of the central angle α radians is $\frac{1}{2} \alpha (r_1 + r_2)(r_1 - r_2)$; or, if a_1, a_2 are arcs bounding that portion, the area $= \frac{1}{2}(a_1 + a_2)(r_1 - r_2)$.

Note. The remainder of the work may be omitted without destroying the integrity of the course.

APPENDIX TO PLANE GEOMETRY.

1. SUPPLEMENTARY THEOREMS IN MENSURATION.

PROPOSITION XIX.

309. Theorem. *If the sides of a triangle are* a, b, c, *and if* $s = \frac{1}{2}(a + b + c)$, $s_1 = s - a$, $s_2 = s - b$, $s_3 = s - c$, *then the area equals* $\sqrt{s \cdot s_1 \cdot s_2 \cdot s_3}$.

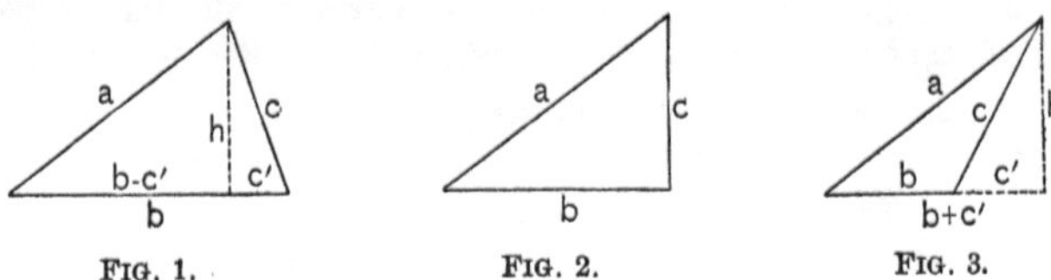

FIG. 1.   FIG. 2.   FIG. 3.

Proof. 1. $a^2 = b^2 + c^2 \mp 2\,bc'$, $2\,bc'$ taking the sign $-$ for Fig. 1, $+$ for Fig. 3, and being 0 for Fig. 2.   § 159

$\therefore c' = \pm\,(b^2 + c^2 - a^2)\,/\,2\,b$, by solving the above equation for c'.   Axs. 2, 7

2. But $h^2 = c^2 - c'^2 = (c + c')\,(c - c')$ § 154
$= [c + (b^2 + c^2 - a^2)\,/\,2\,b]\,[c - (b^2 + c^2 - a^2)\,/\,2\,b]$, by substituting the value of c' given in step 1.

$\therefore h^2 = (2\,bc + b^2 + c^2 - a^2)\,(2\,bc - b^2 - c^2 + a^2)\,/\,4\,b^2$, by removing parentheses and simplifying.

3. $\therefore 4\,b^2h^2 = [(b + c)^2 - a^2]\,[a^2 - (b - c)^2]$, by multiplying by $4\,b^2$ and factoring.

$\therefore 4\,b^2h^2 = (b + c + a)\,(b + c - a)\,(a + b - c)\,(a - b + c)$, by factoring still farther.

226

4. But if $a + b + c = 2\,s$, as given,

   then $b + c - a = 2\,(s - a) = 2\,s_1,$

   and $a - b + c = 2\,(s - b) = 2\,s_2,$

   and $a + b - c = 2\,(s - c) = 2\,s_3.$

5. $\therefore 4\,b^2 h^2 = 2\,s \cdot 2\,s_1 \cdot 2\,s_2 \cdot 2\,s_3.$ Subst. in 3

6. $\therefore \text{area} = \tfrac{1}{2}\,bh = \sqrt{s \cdot s_1 \cdot s_2 \cdot s_3}.$ V, prop. II, cor. 3

NOTE. This is known as Hero's formula for the area of a triangle. Of course a, b, c represent numerical values as explained under V, prop. II, cor. 2.

PROPOSITION XX.

310. Theorem. *The radius, r, of the circle circumscribed about the triangle* abc *of area* t, *equals* abc/4 t.

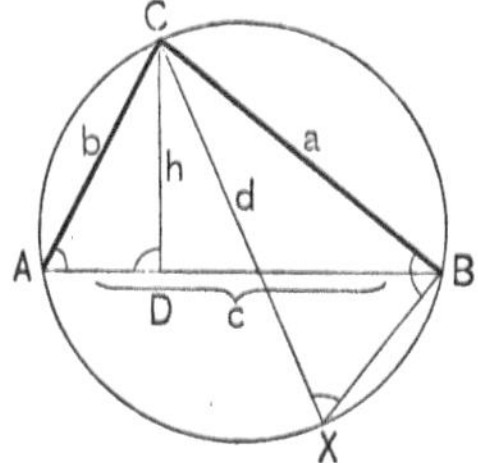

Proof. 1. Suppose $CX = d$ a diameter, CD (or h) $\perp AB$, and BX drawn.

   Then $\triangle ADC \backsim \triangle XBC,$ IV, prop. XVII

   and $\therefore d : a = b : h.$ Why ?

2. $\therefore r = ab / 2\,h.$ IV, prop. I; ax. 7

3. But $\because \tfrac{1}{2}\,hc = t,$ V, prop. II, cor. 3

                  $\therefore r = abc / 4\,t.$ Subst. 3 in 2

NOTE. The value of t can be found by Hero's formula.

Exercise. 477. Find the areas of the triangles with sides (1) 13, 14, 15; (2) 3, 5, 8; (3) 7, 10, 18; (4) a, a, a; (5) 3, 4, 5.

Pʀᴏᴘᴏsɪᴛɪᴏɴ XXI.

311. Theorem. *The product of the diagonals of an inscriptible quadrilateral equals the sum of the products of the opposite sides.*

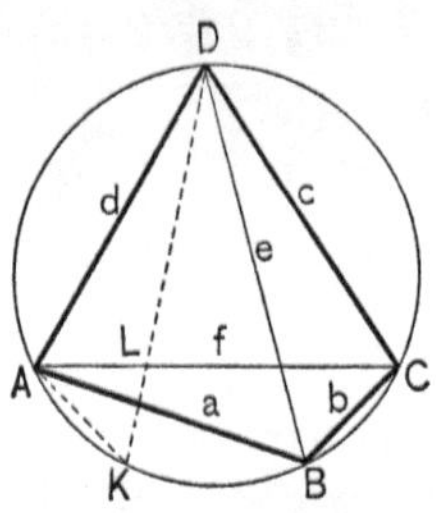

Given $ABCD$, an inscriptible quadrilateral, with sides a, b, c, d, and diagonals e, f.

To prove that $\qquad ef = ac + bd$.

Proof. 1. Let $ABCD$ be inscribed, the sides arranged as in the figure, chord $AK = BC$, and DK drawn cutting AC at L.

Then $\angle ADK = \angle BDC$, and $\angle CAD = \angle CBD$,

and $\qquad \therefore \triangle ALD \backsim \triangle BCD$. $\qquad\qquad$ Why ?

2. Also $\because \angle DCL = \angle DBA$, and $\angle LDC = \angle ADB$,

$\qquad\qquad\qquad$ III, prop. XI, cor. 1 ; ax. 2

$\qquad \therefore \triangle CDL \backsim \triangle BDA$. $\qquad\qquad$ Why ?

3. From 1, $AL : d = b : e$, or $AL = bd/e$; IV, prop. XX

from 2, $LC : c = a : e$, or $LC = ac/e$. IV, prop. XX

4. $\therefore AL + LC$, or AC, or $f = (ac + bd)/e$. $\qquad$ Ax. 2

5. $\qquad\qquad \therefore ef = ac + bd$. $\qquad\qquad$ Ax. 6

Nᴏᴛᴇ. Ptolemy's theorem.

Exercise. 478. Is prop. XXI true when $b =$ zero ?

2. MAXIMA AND MINIMA.

312. Definitions. If a geometric magnitude can, by continuous change, increase until a value is reached at which the magnitude begins to decrease, such value is called a **maximum** value; if it can similarly decrease until a value is reached at which it begins to increase, such value is called a **minimum** value.

In general, a magnitude can have more than one maximum or minimum value, as in the annexed figure where a_1, a_2, a_3 represent maximum, and b_1, b_2, minimum values of the ordinates of F. In the elementary geometry of the line

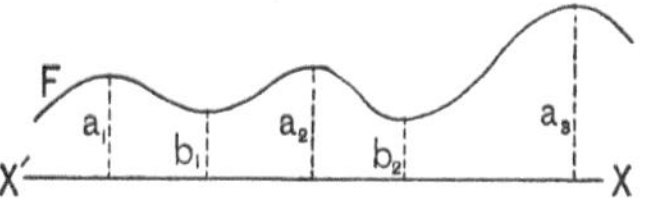

and circle, however, only one maximum or minimum exists, so that the words here mean *greatest* and *least*.

E.g. the maximum chord of a circle is the diameter (III, prop. VIII, cor.), and the minimum chord is spoken of as zero, since zero is the limit which constantly decreasing chords of a circle approach.

A magnitude at its maximum value is called *a maximum;* similarly, *a minimum*. *E.g.* a chord of a circle is a maximum when it is a diameter.

313. Figures having equal perimeters are said to be **isoperimetric.**

Exercises. 479. Draw a line AB, bisect it at M, and take a point X on AM; then show that $AX^2 + XB^2 = 2AM^2 + 2XM^2$, and that this is a minimum when $XM = 0$; hence show that the sum of the squares on the two segments of a given line is a minimum when the segments are equal.

480. Also that $AX \cdot XB = MB^2 - XM^2$, and that this is a maximum when $XM = 0$; that is, that the rectangle of the two segments into which a given line can be divided is a maximum when the given line is bisected.

481. If the diagonals of an inscribed quadrilateral are perpendicular to each other, then the sum of the products of the two opposite sides equals twice the area of the quadrilateral.

Proposition XXII.

314. Theorem. *Of all triangles formed with the same two given sides, that is the maximum whose sides contain a right angle.*

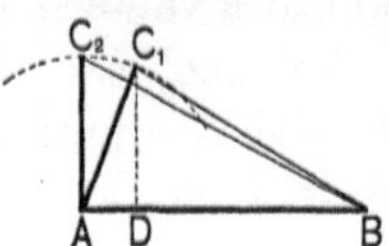

Given   the $\triangle ABC_1$, ABC_2, with $AC_1 = AC_2$, and $AC_2 \perp AB$.

To prove that $\triangle ABC_2 > \triangle ABC_1$.

Proof.   Suppose $C_1D \perp AB$.

Then $AC_1 > DC_1$,   I, prop. XX

and $\therefore$ its equal $AC_2 > DC_1$.

$\therefore \triangle ABC_2 > \triangle ABC_1$,   II, prop. II, cor. 3

since they have the same bases but different altitudes.

Exercises. 482. Find in radians the angle α of a sector of a circle of radius r, such that the number of square units of its area equals the number of linear units of its entire perimeter.

483. Interpret the result of ex. 482 for $r = 2\left(1 + \dfrac{1}{\pi}\right)$. Discuss it for $r \not< 2$. Discuss it for $r < 2\left(1 + \dfrac{1}{\pi}\right)$.

484. In the Sulvasutras, early semi-theological writings of the Hindus, it is said: "Divide the diameter into 15 parts and take away 2; the remainder is approximately the side of the square equal to the circle." From this compute their value of π.

485. On AB describe a semicircle, and in it inscribe the isosceles triangle ABC; on BC and CA describe semicircles opposite the $\triangle ABC$. Show that $\triangle ABC =$ the sum of the two lunes thus formed. (The lunes of Hippocrates.)

486. Six lights are placed regularly on the circumference of a circle of radius 21 ft.; what are the distances of each from each of the others? (To 0.01.)

Proposition XXIII.

315. Theorem. *Of all isoperimetric triangles on the same base the isosceles is the maximum.*

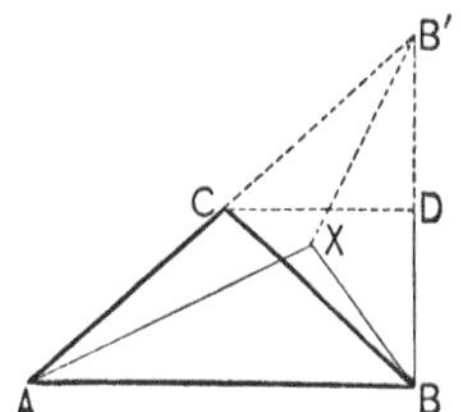

Given　　two isoperimetric $\triangle ABC$ and ABX, $\triangle ABC$ being isosceles, with $AC = BC$.

To prove that　　$\triangle ABC > \triangle ABX$.

Proof.　1. On AC produced, let $CB' = AC$; draw $B'B$, $B'X$; suppose $CD \parallel AB$.

　　　　Then　　$\because AC = CB'$,

　　　　　　$\therefore BD = DB'$.　　　I, prop. XXVII, cor. 2

　2. And　　$\because CB = AC$,

　　　　$\therefore CB = CB'$, and $CD \perp BB'$. Why? Ax. 1

　3.　$\therefore AC + CB = AC + CB' < AX + XB'$.

　　　　　　　　　　Ax. 2; I, prop. VIII

　4.　$\because AX + XB = AC + CB$,　　　　Why?

　　　$\therefore AX + XB < AX + XB'$,

　　and　　$\therefore XB < XB'$.　　　　　Why?

　5. $\therefore X$ and AB lie on the same side of CD,

　　　　　　　　　I, prop. XX, cor. 3

　　and $\therefore \triangle ABC > \triangle ABX$.　　II, prop. II, cor. 3

Corollary. *Of all isoperimetric triangles, that which is equilateral is the maximum.*　(Why?)

Proposition XXIV.

316. Theorem. *Of all triangles having the same base and area, the isosceles has the minimum perimeter.*

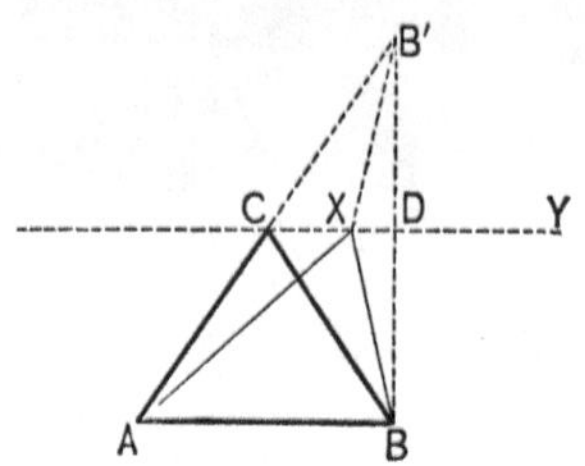

Given the △ ABC and ABX having the same base and area, with $AC = BC$.

To prove that perimeter $ABC <$ perimeter ABX.

Proof. 1. Suppose $CY \parallel AB$; AC produced so that $CB' = AC$; $B'X$ drawn; and $B'B$ drawn cutting CY at D.

Then $\because \triangle ABC = \triangle ABX$,

$\therefore CY$ passes through X.   II, prop. II, cor. 4

2. And $\quad \because AC = CB'$,

$\therefore BD = DB'$.   I, prop. XXVII, cor. 2

3. And $\because \triangle BDC \cong \triangle B'DC$,   I, prop. XII

$\therefore CD \perp BB'$,   Why?

and $\quad \therefore XB = XB'$.   I, prop. XX

4. But $\quad AC + CB' < AX + XB'$,   I, prop. VIII

and $\therefore AC + CB < AX + XB$.

5. $\quad \therefore$ perim. $ABC <$ perim. ABX.   Why?

Corollary. *Of all equal triangles, that which is equilateral has the minimum perimeter.*

For whatever side is taken as the base, the perimeter is less if the other two sides are equal.

Proposition XXV.

317. Theorem.   *If the ends of a line of given length are joined by a straight line, and the area of the figure enclosed is a maximum, it takes the form of a semicircle.*

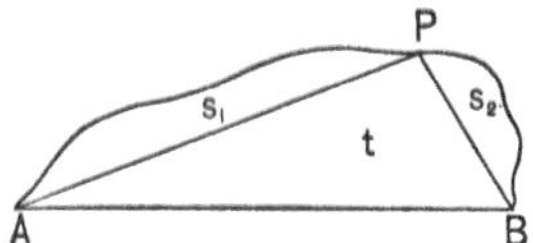

Given   a line *APB* (the curve in the figure), of given length, and *AB* joining its end-points.

To prove   that, if the area of the figure *ABP* is a maximum, *ABP* is a semicircle.

Proof. 1. Let *P* be any point on the line; then joining *A* and *P*, *B* and *P*, let the segments cut off by *AP*, *BP* be called s_1, s_2, and $\triangle ABP$ called t, as in the figure.

Then $\angle P$ is a right angle; for if not, without changing s_1, s_2, the area of t could be increased by making $\angle P$ right.   Prop. XXII

2. But this is impossible if *ABP* is a maximum, and similarly for any other point on *APB*.   Why ?

3. ∴ the area enclosed is a maximum when the line connecting *A* and *B* subtends a right angle at every point on the curve.

Note. It will be seen that examples of maxima or minima involve also the idea of *symmetry* (§ 68). This fact is of value in solving problems in maxima and minima.

Exercise.   **487.** Given the points *A*, *B*, on the same side of line *X'X*, to find on *X'X* a point *P* such that $\angle X'PA = \angle BPX$. Prove that $AP + PB$ is the shortest path from *A* to *X'X* and back to *B*. (Reflected ray of light.)

PROPOSITION XXVI.

318. Theorem. *Of all isoperimetric plane figures the maximum is a circle.*

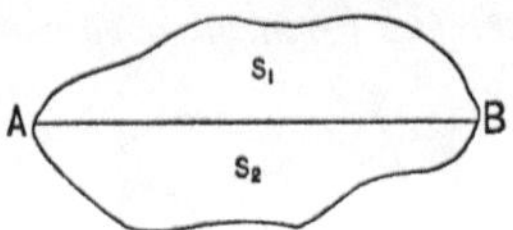

Proof. Suppose A, B points bisecting the given perimeter, AB cutting the figure into two segments, s_1, s_2.

Then s_1, s_2 are maxima when they are semicircles, and AB is a diameter. Why?

PROPOSITION XXVII.

319. Theorem. *Of all equal plane figures the circle has the minimum perimeter.*

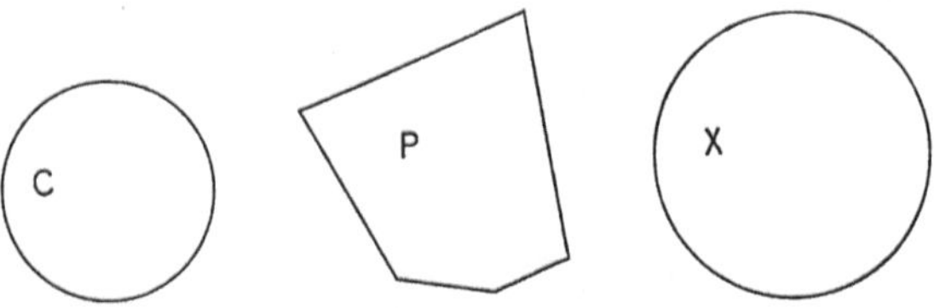

Given circle C = plane figure P.

To prove that circumference $C <$ perimeter P.

Proof. 1. Suppose X a circle of circumference equal to perimeter P.

Then $P < X$, Prop. XXVI

and $\therefore C < X.$ Subst.

2. $\therefore$ circumference $C <$ circumference X, § 307, cor. 4

and $\therefore$ circumference $C <$ perimeter P. Subst.

Proposition XXVIII.

320. Theorem. *A polygon with given sides is a maximum when it is inscriptible.*

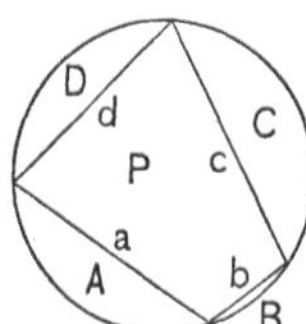 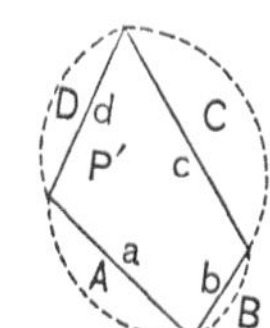

Given   two polygons, P and P', with given sides a, b, c ....., P being inscribed in a circle, and P' not inscriptible.

To prove that $P > P'$.

Proof. 1. Name the circular segments on a, b, ... (opposite P), A, B, ... ; suppose congruent segments constructed on a, b, ... (opposite P').

Then $P + A + B + \cdots > P' + A + B + \cdots$

Prop. XXVI

2. $\therefore P > P'$.   Why ?

Exercises. 488. If the diagonals of a parallelogram are given, its area is a maximum when it is a rhombus.

489. What is the minimum line from a given point to a given line ? Where has this been proved ?

490. Into what two parts must a given number be divided so that the product of those parts shall be a maximum ? (Compare ex. 479.)

491. As a corollary to ex. 479, show that of isoperimetric rectangles the square is the maximum.

492. Find the point in a given straight line such that the tangents drawn from it to a given circle contain the maximum angle.

493. A straight ruler, 1 foot long, slips between the two edges of the floor (the edges making a right angle). Find the position of the ruler when the triangle formed by the edges and ruler is a maximum; also the area of that triangle.

Proposition XXIX.

321. Theorem. *Of all isoperimetric polygons of a given number of sides, the maximum is regular.*

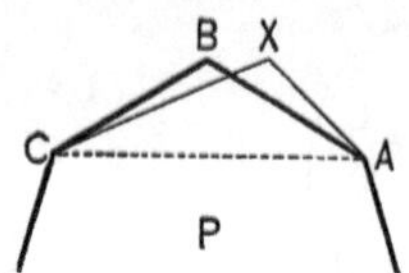

Given P, the maximum polygon of a given perimeter and a given number of sides.

To prove that P is regular.

Proof. 1. Any two adjacent sides, AB, BC, must be equal. For if unequal, as AX, XC, then $\triangle AXC$ could be replaced by $\triangle ABC$, having $AB = BC$, thus enlarging P without changing the perimeter.   But this is impossible because P is a maximum.   Prop. XXIII

2. And hence P is inscriptible because its sides are given.   Prop. XXVIII

3. $\therefore P$ is regular.   V, prop. IX, cor. 2

Exercises.   494. Considering only the relation of space enclosed to amount of wall, what would be the most economical form for the ground plan of a house ?

495. Of all triangles in a given circle, what is the shape of the one having the greatest area ?   Prove it.

496. Through a point of intersection of two circumferences draw the maximum line terminated by the two circumferences.

497. Of all triangles of a given base and area, the isosceles has the greatest vertical angle.

498. Draw the minimum straight line between two non-intersecting circumferences.

Proposition XXX.

322. Theorem. *Of two isoperimetric regular polygons, that having the greater number of sides is the greater.*

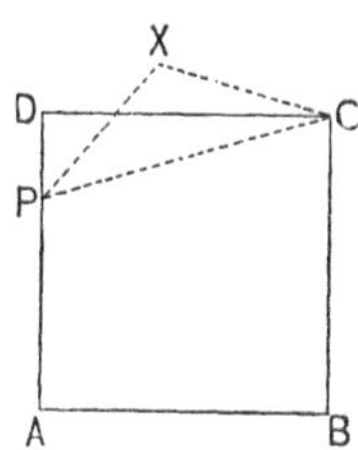

Proof. 1. Let $ABCD$ be a square, P a point on DA, $\triangle PCX$ isoperimetric with $\triangle PCD$ and having $CX = PX$.

Then $\triangle PCX > \triangle PCD$,   Prop. XXIII

and $\therefore$ pentagon $ABCXP > \square ABCD$.   Ax. 4

2. But pentagon $ABCXP$ would, with the same perimeter, be greater if it were regular.   Prop. XXIX

3. $\therefore$ a regular pentagon is greater than an isoperimetric square. Similarly, a regular hexagon would be greater than an isoperimetric regular pentagon, and so on.

Exercises. 499. A cross-section of a bee's cell is a regular hexagon. Show that this is the best form for securing the greatest capacity with a given amount of wax (perimeter).

500. Find the maximum rectangle inscribed in a given semicircle.

501. Find the minimum square inscribed in a given square.

502. Draw the minimum tangent from a variable point in a given line to a given circle.

503. What is the area of the largest triangle that can be inscribed in a circle of radius 5 ?

504. Given a square of area 1. Find the area of an isoperimetric (1) equilateral triangle, (2) regular hexagon, (3) circle.

3. CONCURRENCE AND COLLINEARITY.

PROPOSITION XXXI.

323. Theorem. *If* X, Y, Z *are three points on the sides* a, b, c, *respectively, of a triangle* ABC, *such that the perpendiculars to the sides at these points are concurrent, then*

$$(BX^2 - XC^2) + (CY^2 - YA^2) + (AZ^2 - ZB^2) = 0 ;$$

and conversely.

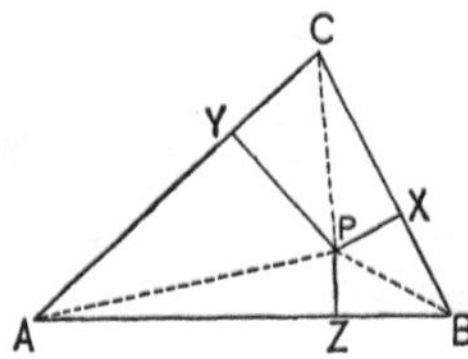

Proof.   Let P be the point of concurrence, and draw PA, PB, PC.

Then $(BX^2 - XC^2) + (CY^2 - YA^2) + (AZ^2 - ZB^2)$
$= PB^2 - PC^2 + PC^2 - PA^2 + PA^2 - PB^2 = 0,$
for $BX^2 - XC^2 = (BP^2 - PX^2) - (PC^2 - PX^2) = BP^2 - PC^2$, and so for the rest.

CONVERSELY: 1. Suppose the $\perp$s from X, Y, to meet at P; and suppose $PZ' \perp c.$

Then as above,
$$(BX^2 - XC^2) + (CY^2 - YA^2) + (AZ'^2 - Z'B^2) = 0.$$

2. But $(BX^2 - XC^2) + (CY^2 - YA^2) + (AZ^2 - ZB^2) = 0,$
and $AZ'^2 - Z'B^2 = AZ^2 - ZB^2.$ Why ?

3. $\therefore AZ'^2 - AZ^2 = Z'B^2 - ZB^2$; but these differences have opposite signs and cannot be equal unless each is zero.

4. $\therefore Z$ must coincide with Z'.

Proposition **XXXII.**

324. Theorem. *If three lines,* x, y, z, *drawn from the vertices of triangle* ABC *to meet* a, b, c *in* X, Y, Z, *are concurrent, then* $\dfrac{AZ}{ZB} \cdot \dfrac{BX}{XC} \cdot \dfrac{CY}{YA} = 1;$ *and conversely.*

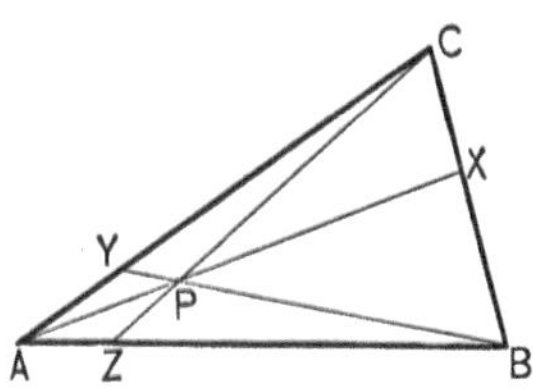

Proof. 1. Let P be the point of concurrence. Then $\because \triangle APC$, PBC have the base PC, they are proportional to their altitudes, and $\therefore$ to AZ, ZB.   Why ?

2. $\therefore \dfrac{AZ}{ZB} = \dfrac{\triangle APC}{\triangle PBC},$

and $\dfrac{BX}{XC} = \dfrac{\triangle BPA}{\triangle APC},$

and $\dfrac{CY}{YA} = \dfrac{\triangle PBC}{\triangle BPA}.$

3. $\therefore \dfrac{AZ}{ZB} \cdot \dfrac{BX}{XC} \cdot \dfrac{CY}{YA} = 1.$ Ax. 6

Conversely : Let CP meet c in Z'; then as above,

4. $\dfrac{AZ'}{Z'B} \cdot \dfrac{BX}{XC} \cdot \dfrac{CY}{YA} = 1.$

5. But $\dfrac{AZ}{ZB} \cdot \dfrac{BX}{XC} \cdot \dfrac{CY}{YA} = 1.$ Given

6. $\therefore \dfrac{AZ'}{Z'B} = \dfrac{AZ}{ZB}.$

7. $\therefore Z'$ must coincide with Z.   IV, prop. XI, cor.

Note. Ceva's theorem.

Proposition XXXIII.

325. Theorem. *If three points,* X, Y, Z, *lying respectively on the three sides* a, b, c *of triangle* ABC, *are collinear, then*

$$\frac{AZ}{ZB} \cdot \frac{BX}{XC} \cdot \frac{CY}{YA} = -1;$$

and conversely.

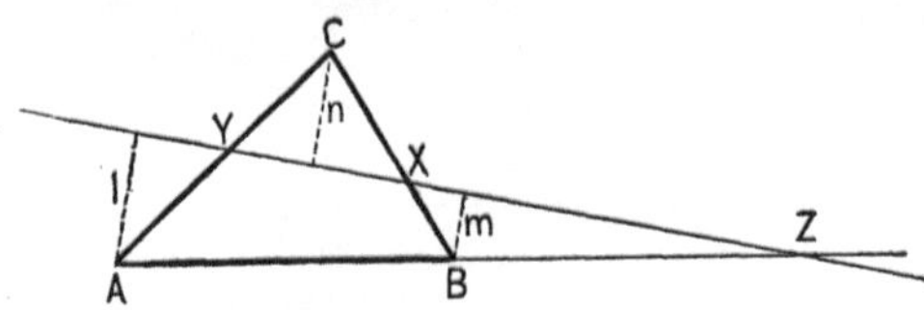

Proof. 1. Let l, m, n be perpendiculars from A, B, C on XY.
Then by similar △, ZB being here negative,

$$\frac{AZ}{ZB} = \frac{l}{-m}.$$

2. And similarly, $\dfrac{BX}{XC} = \dfrac{m}{n}$,

and $\dfrac{CY}{YA} = \dfrac{n}{l}.$

3. $\therefore \dfrac{AZ}{ZB} \cdot \dfrac{BX}{XC} \cdot \dfrac{CY}{YA} = -1.$ Ax. 6

Conversely: Let XY meet AB in Z'; then as above,

4. $\dfrac{AZ'}{Z'B} \cdot \dfrac{BX}{XC} \cdot \dfrac{CY}{YA} = -1.$

5. But $\dfrac{AZ}{ZB} \cdot \dfrac{BX}{XC} \cdot \dfrac{CY}{YA} = -1.$ Given

6. $\therefore \dfrac{AZ'}{Z'B} = \dfrac{AZ}{ZB}.$

7. $\therefore Z'$ must coincide with Z.   IV, prop. XI, cor.

Noᴛᴇ. Menelaus's theorem.

Proposition XXXIV.

326. Theorem. *If a circumference intersects the sides,* a, b, c, *of a triangle* ABC, *in the points* A_1 *and* A_2, B_1 *and* B_2, C_1 *and* C_2, *respectively, then*

$$\frac{AC_1}{C_1B} \cdot \frac{BA_1}{A_1C} \cdot \frac{CB_1}{B_1A} \cdot \frac{AC_2}{C_2B} \cdot \frac{BA_2}{A_2C} \cdot \frac{CB_2}{B_2A} = 1.$$

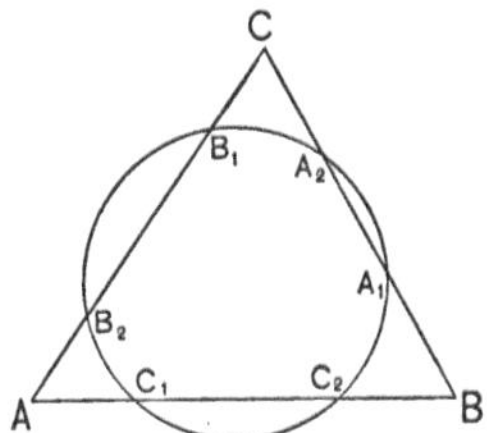

Proof. 1. $\qquad AC_1 \cdot AC_2 = B_1A \cdot B_2A,$ $\qquad$ Why ?

$\qquad$ and $\qquad BA_1 \cdot BA_2 = C_1B \cdot C_2B,$

$\qquad$ and $\qquad CB_1 \cdot CB_2 = A_1C \cdot A_2C.$

2. $\therefore$ by axs. 6 and 7, the above result follows.

Note. This theorem, known as Carnot's theorem, is not a proposition in concurrence or collinearity. It is introduced as leading to the proof of the very celebrated theorem following, one commonly known as the Mystic Hexagram, discovered by Pascal at the age of 16.

The theorem is also easily proved when the triangle is inscribed or circumscribed.

Exercises. 505. By means of Ceva's theorem, prove that the three medians of a triangle are concurrent.

506. Also, that the bisectors of the three interior angles of a triangle are concurrent.

507. Also, that the bisectors of two exterior and of the other interior angles of a triangle are concurrent.

508. Also, that the perpendiculars from the vertices of a triangle to the opposite side are concurrent.

Proposition XXXV.

327. Theorem. *If the opposite sides of an inscribed hexagon intersect, they determine three collinear points.*

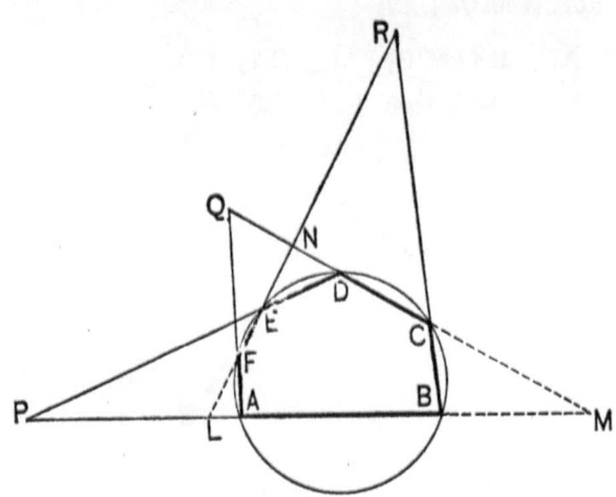

Given an inscribed hexagon, $ABCDEF$, such that BA and DE meet at P, CD and AF at Q, BC and FE at R.

To prove that P, Q, R are collinear.

Proof. 1. Call the $\triangle$ determined by AB, CD, and EF, LMN, as in the figure.

Then from Menelaus's theorem,

$$\frac{LP}{PM} \cdot \frac{MD}{DN} \cdot \frac{NE}{EL} = -1,$$

and $$\frac{MQ}{QN} \cdot \frac{NF}{FL} \cdot \frac{LA}{AM} = -1,$$

and $$\frac{NR}{RL} \cdot \frac{LB}{BM} \cdot \frac{MC}{CN} = -1.$$

2. $\therefore$ By multiplying and recalling Carnot's theorem,

$$\frac{LP}{PM} \cdot \frac{MQ}{QN} \cdot \frac{NR}{RL} = -1.$$

3. $\therefore$ by Menelaus's theorem, P, Q, R are collinear.

MISCELLANEOUS EXERCISES.

509. Show that the following is a special case of prop. **XXXI**: The perpendicular bisectors of the sides of a triangle are concurrent.

510. Also, the perpendiculars from the vertices of a triangle to the opposite sides are concurrent.

511. If three circumferences intersect in pairs, the common chords are concurrent.

512. By means of Menelaus's theorem, prove that the points in which the three bisectors of the exterior angles of a triangle meet the opposite sides are collinear.

513. Also, that the points in which the two bisectors of two interior angles of a triangle and the other exterior angle meet the opposite sides are collinear.

514. The orthocenter, O, of $\triangle ABC$ is determined by the perpendiculars AD, BE. Prove that $AO \cdot OD = BO \cdot OE$.

515. Draw a circle with a central right angle AOB, A and B being on the circumference; bisect $\angle AOB$ by OM, meeting $\overset{\frown}{AB}$ at M; draw $MP \perp OA$; then see if the following is true in general: $\overset{\frown}{AB} = $ chord $AB + PA$. (Consider special cases, $\overset{\frown}{AB} = 120°$, $180°$, $360°$.)

516. Given the base and the vertical angle of a triangle; construct it so that its area shall be a maximum.

517. AB is a diameter of a circle of center O; from any point P on the circumference, PC is drawn perpendicular to AB; from C a perpendicular CE is drawn to OP. Prove that PC is a mean proportional between OA and PE.

518. On side a of $\triangle ABC$, point P is taken such that $\angle PAC = \angle B$. Prove that $CP : CB = AP^2 : AB^2$. Investigate for three cases, $\angle A <, =, > \angle B$.

519. ABC is a triangle right-angled at C; $CD \perp c$. Prove that $AD : DB = CA^2 : BC^2$.

520. If O, O' are the centers of two fixed circles, such that the circumference of O' passes through O, and if a tangent to circumference of O at T cuts circumference of O' at X, Y, then $OX \cdot OY$ is constant. (If the center-line meets the circumference of O' at A, $\triangle XTO \backsim \triangle AYO$.)

521. If O is the orthocenter of triangle ABC, and A', B', C' are the mid-points of a, b, c; M_a, M_b, M_c are the mid-points of AO, BO, CO; P_a, P_b, P_c are the feet of the perpendiculars from A, B, C to a, b, c; prove that A', B', C', M_a, M_b, M_c, P_a, P_b, P_c are concyclic. (The "Nine Points Circle.")

328. NUMERICAL TABLES.

Formulæ of Mensuration. The numbers refer to the pages. Abbreviations : b, base ; h, altitude ; r, radius ; a, area ; c, circumference ; p, perimeter.

Parallelogram, 202,	$a = bh.$	Circle, 217, 224,	$c = 2\pi r.$
Triangle, 202,	$a = \frac{1}{2}bh.$		$a = \pi r^2.$
Trapezoid, 202,	$a = \frac{1}{2}(b + b')h.$	Arc, 223,	$= \alpha \cdot r.$

Most Important Expressions involving π.

$\pi = 3.141593.$	$1/\pi = 0.31830989.$	$180°/\pi = 57°.29578.$
$\pi/4 = 0.785398.$	$\pi^2 = 9.86960440.$	$\pi/180 = 0.01745.$
$\pi/3 = 1.047198.$	$\sqrt{\pi} = 1.77245385.$	Approximate values.
$\frac{4}{3}\pi = 4.188790.$	$1/\sqrt{\pi} = 0.56418958.$	$\pi = \frac{22}{7} = 3\frac{1}{7}, \frac{355}{113}.$

Certain Numerical Results frequently used.

$\sqrt{2} = 1.4142.$	$\sqrt{5} = 2.2361.$	$\sqrt{7} = 2.6458.$
$\sqrt{3} = 1.7321.$	$\sqrt{6} = 2.4495.$	$\sqrt{\frac{1}{2}} = 0.7071.$

329. BIOGRAPHICAL TABLE.

The following table includes only those names mentioned in this work, although numerous others might profitably be considered by the student. The history of geometry may be said to begin in Egypt, the work of Ahmes, copied from a treatise of about 2500 B.C., containing numerous geometric formulæ. The scientific study of the subject did not begin, however, until Thales visited that country, and carried the learning of the Egyptians back to Greece. The period of about four hundred years from Thales to Archimedes may be called the golden age of geometry. The contributions of the latter to the mensuration of the circle and of certain solids practically closed the scientific study of the subject in ancient times. Only a few contributors, such as Hero, Ptolemy, and Menelaus, added anything of importance during the eighteen hundred years which preceded the opening of the seventeenth century. Within the past three hundred years several important propositions and numerous improvements in method have been added, but the great body of elementary plane geometry is quite as Euclid left it. In recent times a new department has been created, known as Modern Geometry, involving an extensive study of loci, collinearity, concurrence, and other subjects beyond the present range of the student's knowledge.

The pronunciations here given are those of the Century Cyclopedia of Names. The first date indicates the year of birth, the second the year of death. All dates are A.D. unless the contrary is indicated by the sign —. The letter *c.* stands for *circa*, about, *b.* for *born*, *d.* for *died*. Numbers after the biographical note refer to pages in this work.

KEY. L. Latin, G. Greek, dim. diminutive, fem. feminine.

a *fat,*	ā *fate,*	ä *far,*	â *fall,*	ȧ *ask,*	ã *fare,*
e *met,*	ē *mete,*	ė *her,*	i *pin,*	ī *pine,*	o *not,*
ŏ *note,*	ö *move,*	ô *nor,*	u *tub,*	ū *mute,*	u̇ *pull.*

ṅ French nasalizing *n.*   ċh German *ch.*

ṣ as in *leisure.*   ṭ as in *nature.*

A single dot under a vowel indicates its abbreviation.

A double dot under a vowel indicates that the vowel approaches the short sound of *u*, as in *put*.

Ahmes (ä′mes). c. − 1700. Egyptian priest. Wrote the oldest extant work on mathematics 221

Anaxagoras (an-aks-ag′o̤-ras). − 499, − 428. Greek philosopher and mathematician 225

Archimedes (är-ki-me′dēz). − 287, − 212. Syracuse, Sicily. The greatest mathematician and physicist of antiquity 87, 221

Aryabhatta (är-yä-bhä′ta). b. 476. One of the earliest Hindu mathematicians. Wrote on Algebra and Geometry 221

Bhaskara (bhäs′ka-ra). 12th cent. Hindu mathematician . . . 104

Brahmagupta (bräh-ma-göp′ta). b. 598. Hindu mathematician. One of the earliest Indian writers 143, 221

Carnot (kär-nō′), Lazare Nicholas Marguerite. 1753, 1823. French physicist and mathematician. Contributed to Modern Geometry 241, 242

Ceulen (koi′len), Ludolph van. 1540, 1610. Dutch geometrician . 221

Ceva (chä′vä), Giovanni. 1648, c. 1737. Italian geometrician, 239, 241

Dase (dä′ze), Zacharias. 1824, 1861. Famous German computer . 221

Euclid (ū′klid). c. − 300. Eminent writer on Geometry in the Alexandrian School, at Alexandria, Egypt. His "Elements," the first scientific text-book on the subject, is still the standard in the schools of England 76, 152, 208

Euler (oi′ler), Leonhard. 1707, 1783. Swiss. One of the greatest mathematicians of modern times 99, 108

Gauss (gous), Karl Friedrich. 1777, 1855. German. One of the greatest mathematicians of modern times 208, 212

Hero (hē′rō) of Alexandria. More properly Heron (hē′ron). c. − 110. Celebrated Greek surveyor and mechanician 221, 227

Hippocrates (hi-pok′ra-tēz) of Chios. b. c. − 470. Author of the first elementary text-book on Geometry 230

Jones (jōnz), William. 1675–1749. English teacher 221

Klein (klīn), Christian Felix. b. 1849. Professor at Göttingen . . 225

Leibnitz (līb′nits), Gottfried Wilhelm. 1646, 1716. Equally celebrated as a philosopher and a mathematician. One of the founders of the science of the Calculus 23, 182

Lindemann (lin′de-man), Ferdinand. b. 1852. German professor 225

Meister (mīs′ter), Albrecht. 1724, 1788. German mathematician 98

Menelaus (men-e-lä′us). c. 100. Greek mathematician and astronomer. One of the early writers on Trigonometry . . 240, 242, 243

Metius (met′ius). Anthonisz, Adriæn. Called Metius from Metz, his birthplace. 1527, 1607 221

TABLE OF ETYMOLOGIES.

Tʜɪs table includes such of the pronunciations and etymologies of the more common terms of Geometry as are of greatest value to the student. The equivalent foreign word is not always given, but rather the primitive root as being more helpful. The pronunciations and etymologies are those of the Century Dictionary. See Biographical Table, p. 372.

Abscissa (ab-sis′ä). L. cut off.

Acute (a-kūt′). L. *acutus*, sharp.

Adjacent (a̤-jä′sent). L. *ad*, to, + *jacere*, lie.

Angle (ang′gl). L. *angulus*, a corner, an angle ; G. *ankylos*, bent.

Antecedent (an-tē-se′dent). L. *ante*, before, + *cedere*, go.

Bisect (bī-sekt′). L. *bi-*, two-, + *sectus*, cut.

Center (sen′tẽr). L. *centrum*, center; G. *kentron*, from *kentein*, to prick.

Centroid (sen′troid). G. *kentron*, center, + *eidos*, form.

Chord (kôrd). G. *chorde*, string.

Circle (sir′kl). L. *circulus*, dim. of *circus* (G. *kirkos*), a ring.

Circumference (sẽr-kum′fe-rens). L. *circum*, around (see Circle), + *ferre*, to bear.

Collinear (ko-lin′ē-är). L. *com-*, together, + *linea*, line.

Commensurable (ko-men′sū-ra-bl). L. *com-*, together, + *mensurare*, measure.

Complement (kom′plē-ment). L. *complementum*, that which fills, from *com-* (intensive) + *plere*, fill.

Concave (kon′-kăv). L. *com-* (intensive) + *cavus*, hollow.

Concentric (kon-sen′trik). L. *con-*, together, + *centrum*, center.

Concurrent (kon-kur′ent). L. *con-*, together, + *currere*, run.

Concyclic (kon-sik′lik). L. *con-*, together, + *cyclicus*, from G. *kyklikos*, from *kyklos*, a circle, related to *kyliein*, roll (compare Cylinder).

Congruent (kong′grö-ent). L. *congruere*, to agree.

Consequent (kon′sē-kwent). L. *con-*, together, + *sequi*, follow.

Constant (kon′stänt). L. *con-*, together, + *stare*, stand.

Converse (kon′vẽrs). L. *con-*, together, + *vertere*, turn.

Convex (kon′veks). L. *convexus*, vaulted, from *con-*, together, + *vehere*, carry.

Corollary (kor′o-lä-ri). L. *corollarium*, a gift, money paid for a garland of flowers, from *corolla*, dim. of *corona*, a crown.

Cylinder (sil′in-dẽr). G. *kylindros*, from *kyliein*, roll.

Decagon (dek'a̯-gon). G. *deka*, ten, + *gonia*, an angle.

Degree (dē-grē'). L. *de*, down, + *gradus*, step.

Diagonal (dĭ-ag'o̯-na̯l). G. *dia*, through, + *gonia*, a corner, an angle.

Diameter (di-am'e-tėr). G. *dia*, through, + *metron*, a measure.

Dihedral (dĭ-hē'dra̯l). G. *di-*, two, + *hedra*, a seat.

Dimension (di-men'sho̯n). L. *dis-*, apart, + *metiri*, measure. See Measure.

Directrix (di-rek'triks). L. fem. of *director*, from *directus*, direct.

Dodecahedron (dō''dek-a̯-hē'dro̯n). G. *dodeka*, twelve, + *hedra*, a seat.

Equal (ē'kwa̯l). L. *æqualis*, equal, from *æquus*, plain.

Equiangular (ē-kwi-ang'gū-lẽr). L. *æquus*, equal, + *angulus*, angle.

Equilateral (ē-kwi-lat'e̯-ra̯l). L. *æquus*, equal, + *latus*, side.

Equivalent (ē-kwiv'a̯-le̯nt). L. *æquus*, equal, + *valere*, be strong.

Escribed (es-krībd'). L. *e*, out, + *scribere*, write.

Excess (ek-ses'). L. *ex*, out, + *cedere*, go ; *i.e.* to pass beyond.

Frustum (frus'tum). L. a piece.

Generatrix (jen'e̯-rā-triks). L. fem. of *generator*, from *generare*, beget, from *genus*, a race.

Geometry (jē̯-om'e-tri). G. *ge*, the earth, + *metron*, a measure.

-gon, a termination, G. *gonia*, an angle.

Harmonic (här-mon'ĭk). G. *harmonia*, a concord, related to *harmos*, a joining. A line divided internally and externally in the ratio 2 : 1, is cut into segments representing 1, ⅔, ½. Pythagoras first discovered that a vibrating string stopped at half its length gave the octave of the original note, and stopped at two-thirds of its length gave the fifth. Hence the expression "harmonic division" of a line.

Hemisphere (hem'i-sfēr). G. *hemi-*, half, + *sphaira*, a sphere.

-hedron, a termination, G. *hedra*, a seat.

Hepta-, in combination, G. seven.

Hexa-, in combination, G. six.

Hexagram (hek'sa̯-gram). G. *hex*, six, + *gramma*, a line.

Hypotenuse (hī-pot'e-nūs). G. *hypo*, under, + *teinein*, stretch.

Inclination (in-kli-nā'sho̯n). L. *in*, on, + *clinare*, lean.

Incommensurable (in-ko̯-men'sū-ra̯-bl). L. *in-*, not, + *com-*, together, + *mensurare*, measure.

Infinity (in-fin'ĭ-ti). L. *in-*, not, + *finitus*, bounded.

Inscribed (in-skrībd'). L. *in*, in, + *scribere*, write.

Isosceles (ī-sos'e-lēz). G. *isos*, equal, + *skelos*, leg.

Lateral (lat'e̯-ra̯l). L. *latus*, a side.

Locus (lō'kus). L. a place. Compare locality.

Lune (lūn). L. *luna*, the moon.

Major (mā'jo̯r). L. greater, comparative of *magnus*, great.

Maximum (mak'si-mum). L. greatest, superlative of *magnus*, great.

Mean (mēn). L. *medius*, middle.

Measure (mezh'ûr). L. *mensura,* a measuring. See Dimension.

Median (mḗ'di-ạn). See Mean.

Mensuration (men-sụ̄-rā'shọn). See Measure.

Minimum (min'i-mum). L. least.

Minor (mī'nọr). L. less.

Nappe (nap). French, a cloth, sheet, surface.

Oblique (ọb-lḗk' or ọb-lĭk'). L. *ob,* befoŕe, + *liquis,* slanting.

Obtuse (ọb-tūs'). L. *obtusus,* blunt, from *ob,* upon, + *tundere,* strike.

Octo-, octa-, in combination, L. and G., eight.

Opposite (op'ọ-zit). L. *ob,* before, against, + *ponere,* put, set.

Ordinate (ôr'di-nạt). L. *ordo (ordin-*), a row.

Orthocenter (ôr'thọ-sen-tėr). G. *ortho-,* straight, + *kentron,* center.

Orthogonal (ôr-thog'ọ-nạl). G. *orthos,* right, + *gonia,* an angle.

Parallel (par'ạ-lel). G. *para,* beside, + *allelon,* one another.

Parallelepiped (par-ạ-lel-e-pip'ed or -pĭ'ped). Gr. *parallelos,* parallel, + *epipedon,* a plane surface, from *epi,* on, + *pedon,* ground.

Parallelogram (par-ạ-lel'ọ-gram). G. *parallelos,* parallel, + *gramma,* line.

Pedal (ped'al or pḗ'dạl). L. *pedalis,* pertaining to the foot, from *pes (ped-*), foot.

Pencil (pen'sil). L. *penicillum,* a painters' pencil, a brush.

Perigon (per'i-gon). G. *peri,* around, + *gonia,* a corner, angle.

Perimeter (pḗ-rim'e-tėr). G. *peri,* around, + *metron,* measure.

Perpendicular (pėr-pẹn-dik ṇ-lạr).

L. *perpendiculum,* a plumb-line, from *per,* through, + *pendere,* hang.

Perspective (pėr-spek'tiv). L. *per,* through, + *specere,* see.

π (pī). Initial of G. *periphereia,* periphery, circumference.

Pole (pōl). G. *polos,* a pivot, hinge, axis, pole.

Polygon (pol'i-gon). G. *polys,* many, + *gonia,* corner, angle.

Polyhedron. (pol-i-hḗ'drọn) G. *polys,* many, + *hedra,* seat.

Postulate (pos'tụ̄-lāt). L. *postulatum,* a demand, from *poscere,* ask.

Prism (prizm). G. *prisma,* something sawed, from *priein,* saw.

Prismatoid (priz'mạ-toid). G. *prisma (t-*), + *eidos,* form.

Projection (prọ-jek'shọn). L. *pro,* forth, + *jacere,* throw.

Pyramid (pir'ạ-mid). G. *pyramis,* a pyramid, perhaps from Egyptian *pir-em-us,* the slanting edge of a pyramid.

Quadrant (kwod'rạnt). L. *quadran(t-)s,* a fourth part. See Quadrilateral.

Quadrilateral (kwod-ri-lat'ẹ-rạl). L. *quatluor (quadri-*), four, + *latus, (later-*), side.

Radius (rā'di-us). L. rod, spoke of a wheel.

Ratio (rā'shiō). L. a reckoning, calculation, from *reri,* think, estimate.

Reciprocal (rẹ-sip'rọ-kạl). L. *reciprocus,* returning, from *re-,* back, and *pro,* forward, with two adjective terminations.

Rectangle (rek'tang-gl). L. *rectus,*

right, + *angulus*, an angle. See
Angle.

Rectilinear (rek-ti-lin′ẹ-är). L. *rectus*, right, + *linea*, a line.

Reflex (rē′fleks or rẹ-fleks′). L. *re-*, back, + *flectere*, bend.

Regular (reg′ụ-lär). L. *regula*, a rule, from *regere*, rule, govern.

Rhombus (rom′bus). G. *rhombos*, a spinning top.

Scalene (skạ-lēn′). G. *skalenos*, uneven, unequal ; related to *skellos*, crooked-legged.

Secant (sē′kant). L. *secare*, cut, as also Sector, Section, Segment.

Segment (seg′ment). See Secant.

Semicircle (sem′i-sèr-kl). L. *semi-*, half, + *circulus*, circle.

Similar (sim′i-lär). L. *similis*, like.

Solid (sol′id). L. *solidus*, firm, compact.

Sphere (sfēr). G. *sphaira*, a ball.

Square (skwãr). L. *quadra*, a square, from *quattuor*, four.

Straight (strāt). Anglo - Saxon, *streht*, from *streccan*, stretch.

Subtend (sub-tend′). L. *sub*, under, + *tendere*, stretch.

Successive (suk-ses′iv). L. *sub*, under, + *cedere*, go.

Sum (sum). L. *summa*, highest part. Compare Summit.

Superposition (sū′pèr-pọ-zish′ọn). L. *super*, over, + *ponere*, lay.

Supplement (sup′lẹ-ment). L. *sub*, under, + *plere*, fill ; to fill up.

Surface (sèr-fās). L. *superficies*, surface, from *super*, above, + *facies*, form, figure, face.

Symbol (sim′bọl). G. *symbolos*, a sign by which one infers something, from *syn*, together, + *ballein*, put.

Tangent (tan′jẹnt). L. *tangere*, touch.

Tetrahedron (tet-ra-hē′drọn). G. *tetra-*, four, + *hedra*, seat.

Theorem (thē′ō-rem). G. *theorema*, a sight, a principle contemplated.

Transversal (tráns-vèr′sạl). L. *trans*, across, + *vertere*, turn.

Trapezium (trạ-pē′zi-um). G. *trapezion*, a table, dim. of *trapeza*, a table, from *tetra*, four, + *pous*, foot.

Trapezoid (trạ-pē′zoid). G. *trapeza*, table, + *eidos*, form.

Tri-, in composition, L. *tres (tri-)*, G. *treis (tri-)*, three. See Secant, -hedron, Angle, for meaning of trisect, trihedral, triangle.

Truncate (trung′kāt). L. *truncare*, cut off, from Old L. *troncus*, cut off, mutilated.

Unique (ụ-nēk′). L. *unicus*, from *unus*, one.

Vertex (vèr′teks). L. *vertere*, turn.

Zone (zōn). G. *zone*, a girdle, belt.

9 781418 165888